Collective Dynamics in Complex Networks of Noisy Phase Oscillators

Towards Models of Neuronal Network Dynamics

Dissertation
zur Erlangung des akademischen Grades

doctor rerum naturalium
(Dr. rer. nat.)

im Fach Physik

eingereicht an der
Mathematisch-Naturwissenschaftlichen Fakultät
der Humboldt-Universität zu Berlin

von
M.Sc. Bernard Sonnenschein

Präsidentin/Präsident der Humboldt-Universität zu Berlin
Prof. Dr.-Ing. Dr. Sabine Kunst

Dekanin/Dekan der Mathematisch-Naturwissenschaftlichen Fakultät
Prof. Dr. Elmar Kulke

Gutachter/-innen: 1. Prof. Dr. Lutz Schimansky-Geier
2. Prof. Dr. Harald Engel
3. Prof. Dr. Pere Colet

Tag der mündlichen Prüfung: 16. Dezember 2015

Zugl.: Diss., Univ., Humboldt-Universität zu Berlin, 2015

Bibliographic information published by the Deutsche Nationalbibliothek

The Deutsche Nationalbibliothek lists this publication in the Deutsche Nationalbibliografie; detailed bibliographic data are available on the Internet at http://dnb.d-nb.de .

ISBN 978-3-8325-4375-4

Logos Verlag Berlin GmbH
Comeniushof, Gubener Str. 47,
10243 Berlin
Tel.: +49 (0)30 42 85 10 90
Fax: +49 (0)30 42 85 10 92
INTERNET: http://www.logos-verlag.de

Zusammenfassung

Wechselwirkende oszillierende Elemente neigen zu gegenseitiger Synchronisation. Unklarer ist das kollektive Verhalten, wenn die Wechselwirkungen nicht uniform sind oder wenn die Elemente durch Rauschen beeinflusst werden. In der vorliegenden Dissertation gehen wir der Frage nach, was für Zustände in diesen Fällen möglich sind und unter genau welchen Bedingungen diese auftreten. Unsere Arbeit wird von neurowissenschaftlichen Themen motiviert, jedoch sind die erhaltenen Ergebnisse von allgemeinerer Natur und können weitere Anwendungen ermöglichen. Wir charakterisieren die Elemente durch ihre individuellen Eigenschaften und verwenden diesen Ansatz bei Populationen von verrauschten Kuramoto Phasen-Oszillatoren. Ohne Korrelationen zwischen der Anzahl der Verbindungen (Grade) lassen sich komplexe Netzwerke gröber betrachten. Dies ermöglicht uns die Kopplungsstärken abzuschätzen, die den Übergang von Inkohärenz zu partieller Synchronität markieren. Wenn die Grade und die Eigenfrequenzen der Oszillatoren nicht voneinander abhängen, gelangen wir zu früheren Ergebnissen, in denen nun das Verhältnis der ersten beiden Momente der Gradverteilung als Skalierungsfaktor eingeht. Das Erreichen partieller Synchronität wird durch Rauschen und Frequenzunterschiede erschwert, jedoch durch Hinzufügen von Verbindungen erleichtert. Letzteres ist besonders ausgeprägt im Kleine-Welt-Netzwerk Regime. Ein Vergleich zwischen theoretischen und numerischen Ergebnissen zeigt gute Übereinstimmungen für dichte Kleine-Welt-Netzwerke. Wir geben weitere numerische Hinweise, dass die Mittlere-Feld-Näherung gültig ist in ausreichend dichten Netzwerken. Bestimmte positive Korrelationen zwischen dem Grad und der Streuung in den Eigenfrequenzen ergeben eine Maskierung der Grad Heterogenität. So ist es möglich, dass Netzwerke mit gleichen Kopplungsstärken synchronisiert werden können, obwohl sie strukturell verschieden voneinander sind. Mit Rauschen müssen die Korrelationen stärker sein und die Form der Frequenzverteilung muss berücksichtigt werden, um eben diesen Maskierungseffekt zu erhalten. Wenn Rauschen die einzige Form von Unordnung darstellt, dann können wir unter der Annahme, dass die Phasen der Oszillatoren einer zeitabhängigen Gauß-Verteilung folgen, die zeitliche Entwicklung des Synchronisationsprozesses explizit herleiten. Wir zeigen, dass dieses Ergebnis besonders im inkohärenten und im hinreichend synchronen Bereich eine gute Näherung darstellt. Dann kombinieren wir die Gauß'sche Theorie mit der Netzwerkvergröberung, um von Rauschen getriebene aktive Rotatoren zu untersuchen, die anregbares Verhalten zeigen können. Wir konzentrieren uns auf Zufallsnetzwerke, die zwei verschiedene Grade enthalten. Dabei sehen wir, dass im selbstoszillierenden Fall der Synchronisationsübergang wieder vom Verhältnis der ersten beiden Momente der Gradverteilung abhängt. Im Gegensatz dazu hängt die Synchronisierbarkeit im anregbaren Fall nicht-monoton von der Grad Heterogenität ab. Der Wechsel von Inkohärenz in den Ruhezustand oder in die partielle Synchronität gelingt am einfachsten bei einer moderaten Grad Heterogenität. Indem wir selbstoszillierende und anregbare Elemente kombinieren, können wir beobachten, dass das Netzwerk nur dann als Ganzes anregbar ist, wenn die anregbaren Elemente eine geringere Kopplungsstärke haben als die selbstoszillierenden Elemente. Um die Rolle von exzitatorischen und inhibitorischen Verbindungen zu untersuchen, betrachten wir außerdem zwei miteinander global gekoppelte Populationen mit gemischten positiven und negativen Kopplungsstärken. Durch die

Gauß'sche Näherung können wir umfangreiche Phasendiagramme ableiten. Neben dem inkohärenten und dem herkömmlichen synchronen Zustand können Zustände von Uneinigkeit gefunden werden, in denen sich die Population in zwei Gruppen teilt. Wenn die Phasenverzögerung zwischen ihnen weder Null noch maximal ist, tritt ein Drift auf, so dass die Frequenz der gesamten Population verschieden ist von der Frequenz der einzelnen Elemente. Abschließend diskutieren wir die Verbindung zwischen Synchronisation und gemeinsamer Frequenz in einem ähnlichen Modell und entwickeln erste Schritte hin zu einer Erweiterung der Gauß'schen Theorie.

Abstract

Interacting oscillatory units tend to mutually synchronize. If the interactions are non-uniform or if the units are affected by noise, the collective behavior remains elusive. What kind of dynamics can emerge, and under which conditions? We shall address this question in the present thesis. Our work is motivated by neuroscientific topics, yet the results obtained are rather general and may enable further applications. We characterize the units by their individual quantities, and apply this framework to populations of noisy Kuramoto phase oscillators. Without correlations between the number of connections (degrees), one can coarse-grain complex networks to estimate the coupling strengths which mark the transition from incoherence to partial synchrony. If the degrees and the natural frequencies of the oscillators do not depend on each other, we recover previous results rescaled by the ratio of the first two moments of the degree distribution. Achieving partial synchrony is impeded by noise and frequency mismatch, but facilitated by adding connections. The latter is particularly pronounced in the small-world network regime. We compare theoretical with simulation results in dense small-world networks and find good agreement. We give numerical evidence that the mean-field approach is valid, if the network is sufficiently dense. Studying correlations between the degrees and the dispersion in the natural frequencies, we find that certain positive correlations lead to a masking of the degree heterogeneity such that different networks synchronize at the same coupling strengths. With noise, the correlations not only have to be stronger, but also the shape of the frequency distribution has to be included in order to recover the masking effect. Further progress can be made by assuming that the oscillators' phases follow a time-dependent Gaussian distribution. When the only source of disorder comes from noise, we use this approximation to derive the time-dependent level of synchronization. The theory is shown to be valid in the incoherent and sufficiently synchronous regime. Next we combine the Gaussian theory with the network coarse-graining to study coupled noise-driven active rotators, which can perform excitable dynamics. We focus on random networks containing two different degrees. Thereby we find that in the self-oscillatory regime the synchronization transition is again governed by the ratio of the first two moments of the degree distribution. In contrast, the synchronizability in the excitable regime depends non-monotonically on the degree heterogeneity. As a consequence, switching from incoherence to resting or partial synchrony is most easily achieved at a moderate degree heterogeneity. Intermingling self-oscillatory and excitable units, we further observe that the network as a whole is excitable, only if the excitable units have a smaller coupling strength than the self-oscillatory ones. To approach the role of excitatory and inhibitory connections, we study two mutually globally coupled populations with mixed positive and negative coupling strengths. By virtue of the Gaussian approximation we are able to derive the rich phase diagrams. Besides incoherence and the traditional synchronous state, discordant states are found, in which the population splits into two clusters. If the phase lag between them is neither zero nor maximal, a drift emerges such that the common frequency of the whole population differs from the frequency of individual units. Finally, we discuss the connection between synchronization and common frequency in a similar model, and develop first steps towards an extension of the Gaussian theory.

Contents

1. Introduction

This thesis is concerned with the collective dynamics taking place in populations of oscillatory units. The collective behavior we study here is mostly described by the phenomenon of mutual synchronization, which critically relies on interactions between individual elements. In the presence of noise and non-uniform interactions, the collective behavior remains particularly elusive. It is this field of research where we aim to contribute.

Mutual synchronization is understood as the adjustment of individual rhythms, which gives rise to collective oscillations. This mechanism is ubiquitously observed in nature, from planets in the universe to electrons in superconductors, from cells in the heart to neurons in the brain [1,2].

Neuroscientific concepts shall inspire many of the aspects that we investigate in the current thesis. In this respect we would like to mention that neural synchronicity with its alterations is considered to be a mechanism underlying conscious perception [3,4] and neuropsychiatric disorders, such as schizophrenia, autism and epilepsy [5,6].

Indeed, consisting of $\sim 10^{11}$ neurons and $\sim 10^{14}$ synapses, the human brain is a prime example for a highly complex network with a multitude of emergent behaviors [7–9]. Neurons communicate with each other via excursions of their membrane potential, so-called spikes, that are transmitted through the axons and finally to synapses that couple with the dendrites of other neurons. The precise neural code is however still not found [10]. Similarly, the network of anatomical connections linking the neurons in the human brain is largely unknown; even for other mammalian species there are only some databases of large-scale anatomical connection patterns [11].

In the face of the aforementioned gap of knowledge, multifaceted modeling attempts must be sought to find new insights. From a physicist's point of view, one must cope with an indefinite number of variations and special cases (seemingly inherent to biology). The physicist is reminded of the wave-particle duality when learning that the information processing in the brain happens on the level of single neurons but also on the level of large cell assemblies, and that the neural code at least combines imprecise firing rates with exact spiking times [10,11].

The methods we apply here are rooted in statistical physics, dynamical systems theory and stochastic processes. In particular, notions from the theory of liquids and plasmas appear. It is therefore appropriate to speak of a synergetic approach to brain activity [8,9,12].

Throughout our work, the idea is to separate the whole population into subpopulations that comprise oscillators with the same individual quantities. Mean-field descriptions shall be gained in this way. Let us introduce the vectors σ_i, $i = 1, \ldots, N$ to collect all individual quantities. For an example aimed at modeling neuronal network dynamics, those vectors may have the following components:

$$\sigma_i = \begin{pmatrix} \text{Number of dendrites of neuron } i \\ \text{Length of axon of neuron } i \\ \vdots \\ \text{Number of excitatory synapses of neuron } i \\ \text{Number of inhibitory synapses of neuron } i \\ \text{Number of gap junctions of neuron } i \\ \vdots \\ \text{Excitation threshold of neuron } i \\ \text{Firing rate of neuron } i \\ \vdots \end{pmatrix}. \tag{1.1}$$

The grouping according to individual quantities is in general an approximation, since the properties of the oscillatory units are not necessarily localized at the nodes of the network. To put it differently, once there are correlations between the oscillators' properties, the set of vectors (1.1) does not fully characterize the network, because the correlations themselves need to be taken into account.

In this thesis, we adopt a phase description of the neuronal dynamics, which enables analytical studies of synchronization processes, as we briefly explain now. With regard to biological oscillators the idea of the phase description goes back to Winfree (1967). He realized that many biological oscillators can be assumed to be weakly coupled to a good approximation. Then no oscillator is ever perturbed far away from its limit cycle and therefore amplitudes can be considered as fixed [13]. Specifically, using this weak-coupling assumption, any population of self-sustained limit-cycle oscillators can be reduced to a set of N coupled ordinary differential equations for the phase variables ϕ_i:

$$\dot{\phi}_i(t) = \omega_i + Q(\phi_i)\frac{K}{N}\sum_{j=1}^{N} P(\phi_j),\ i = 1, \ldots, N. \tag{1.2}$$

This is the so-called Winfree model. The constant K denotes the coupling strength, and heterogeneity among the oscillators is included through the natural frequencies ω_i. The $Q(\phi_i)$ and $P(\phi_j)$ are phase response and influence functions, respectively, which need to be specified. Kuramoto (1975, 1984) showed through time-averaging in Eq. (1.2) that the two functions $Q(\phi_i)$ and $P(\phi_j)$ can be merged into one interaction function for the phase *differences*, $\Gamma_{ij}(\phi_j - \phi_i)$, if the natural frequencies ω_i are additionally assumed to be nearly identical [14, 15]. Searching for a solvable model, Kuramoto restricted the interaction function to the first Fourier mode, $\Gamma_{ij}(\phi_j - \phi_i) \equiv \sin(\phi_j - \phi_i)$. This choice is now famously known as the Kuramoto model (for reviews see [16, 17] and for overviews about the potential neuroscientific relevance see [18, 19]).

With the classical Kuramoto order parameters [mean-field amplitude $r(t)$ and mean phase $\Theta(t)$],

$$r(t)\mathrm{e}^{i\Theta(t)} := \frac{1}{N}\sum_{j=1}^{N} \mathrm{e}^{i\phi_j(t)}, \tag{1.3}$$

the Kuramoto model can be written as

$$\dot{\phi}_i(t) = \omega_i + r(t)K \sin\left[\Theta(t) - \phi_i(t)\right]. \tag{1.4}$$

In this form the mean-field character becomes evident, because in effect one has N oscillators that are coupled only through the global mean field. One can ask to what extent neurons can be described as (phase) oscillators. In particular strongly irregular dynamics question the underlying assumptions [20, 21]. In general, neurons should be rather described as excitable elements that are perturbed by noise [22]. We will therefore investigate extensions of the Kuramoto model towards excitable dynamics.

The thesis is organized as follows. After the introduction we continue with Chap. 2, which introduces the basic Kuramoto model studied henceforth. We explain how to coarse-grain random networks that are uncorrelated and undirected. We use this approximation to derive a nonlinear Fokker-Planck equation for the one-oscillator probability density. On this basis we perform a linear stability analysis of the incoherent state in order to derive a formula for the onset of synchronization. The analytical results are tested for several examples and against numerical simulations. In Chap. 3 we explain the Gaussian approximation which allows to find drastic dimensionality reductions. For the simplest case of identical noise-perturbed Kuramoto oscillators, it is possible to derive a full time-dependent solution. We test the validity of the Gaussian approximation by comparing with numerical experiments and exact results. Qualitatively new behavior is introduced in Chap. 4 when we study noise-driven active rotators that can perform excitable dynamics. Combining the coarse-graining of the network with the Gaussian approximation allows us to perform thorough bifurcation analyses of reduced systems. In Chap. 5 we return to the stochastic Kuramoto model, but this time we consider asymmetric interactions and we allow for both positive and negative coupling strengths. New collective states are observed thereby, which we are able to understand in the framework of the Gaussian approximation. In Chap. 6 we consider the Kuramoto-Sakaguchi model, the characteristic feature of which is an explicit phase lag in the sinusoidal coupling function. First steps towards an extension of the Gaussian theory are presented, which can cope with the inaccuracies that are caused by the phase lag.

2. Onset of synchronization in networks of noisy phase oscillators

The onset of synchronization is a very distinct event. Suddenly, oscillatory units start to adjust their rhythms, partially agreeing on a new rhythm. This phenomenon is found pretty much through all the sciences, with particular relevance in the neurosciences, as we have briefly elaborated already in the introduction. It is believed that the brain operates preferably in a range close enough to the border between incoherence and synchronicity. This becomes clear with respect to certain dysfunctions. For instance, epileptic seizures [23] and Parkinson's disease [24] result from hypersynchronous neuronal activity, whereas schizophrenia [25,26], autism and Alzheimer's disease [27] seem to be linked with a lack of neural synchrony. For a review on the role of synchrony in brain disorders, see Ref. [28]. The study of synchronization transitions is well-established, but there are still many open questions concerning the conditions of synchronization in presence of temporal fluctuations and non-uniform connectivity. Our investigations in this direction are based on a variant of the Kuramoto model [14,15]:

$$\dot{\phi}_i(t) = \omega_i + \frac{K}{N}\sum_{j=1}^{N} A_{ij}\sin\left[\phi_j(t) - \phi_i(t)\right] + \xi_i(t),\ i = 1,\dots,N\ , \tag{2.1}$$

with the number of oscillators N, the phase $\phi_i(t)$ at time t, and natural frequency ω_i of oscillator i. The coupling strength is denoted by K and A_{ij} are the elements of the adjacency matrix. If the nodes i and j are not connected, then $A_{ij} = 0$, otherwise $A_{ij} = 1$. In general one has to distinguish between in-degree and out-degree of node i, defined as the number of edges pointing to or emanating from node i, respectively [29]. We consider undirected networks here, then the adjacency matrix is symmetric and the in-degrees are equal to the out-degrees. Hence, we refer only to the degree k_i of node i, which can be obtained from the adjacency matrix by summing up:

$$k_i = \sum_{j=1}^{N} A_{ij}. \tag{2.2}$$

We think of the natural frequencies and individual degrees as being distributed according to a joint probability density $P(\omega, k)$. In Sec. A we touch on directed networks.

The division by N in the coupling term of Eq. (2.1) is not always the correct normalization for complex networks, because it might not lead necessarily to an intensive coupling term. This is the case, if the degrees do not scale linearly with the system size. Then one has to take a different normalization, e.g. the maximum degree [30].

The functions $\xi_i(t)$, $i = 1, \ldots, N$, stand for sources of independent white noise processes that act on the natural frequencies as a stochastic force. They can be regarded as an aggregation of various stochastic processes, such as the variability in the release of neurotransmitters or the quasi-random synaptic inputs from other neurons [22]. The noise terms satisfy

$$\begin{aligned} \langle \xi_i(t) \rangle &= 0, \\ \langle \xi_i(t)\xi_j(t') \rangle &= 2D\delta_{ij}\delta(t-t') \,. \end{aligned} \tag{2.3}$$

Hence, a single nonnegative parameter D scales the noise intensity and we assume that it is independent of the system size. The angular brackets denote an average over different realizations of the noise.

Defining the proper phase for stochastic oscillators is in fact very non-trivial [31–39] . We are not touching this problem in the present work, since we add the noise after the phase reduction has been done. Thus, what we consider here are noise-perturbed phase evolutions of otherwise deterministic oscillators.

2.1. Coarse-graining uncorrelated random networks

In the past, the study of critical dynamics on complex networks benefitted from coarse-graining methods, see for instance Refs. [40–44]. The idea is to adopt a combinatorial point of view in order to mimic the complex network by a weighted fully connected network with a new adjacency matrix $\tilde{A}_{ij}$ which should resemble the original structure defined by the given set of degrees k_i, $i = 1, ..., N$. This can be achieved by requiring that the approximating weights $\tilde{A}_{ij}$ conserve all the degrees of the original network [cf. Eq. (2.2)], i.e.

$$k_i \stackrel{!}{=} \sum_{j=1}^{N} \tilde{A}_{ij} \quad \text{and} \quad k_j \stackrel{!}{=} \sum_{i=1}^{N} \tilde{A}_{ij}, \tag{2.4}$$

Here we consider the simplest case of undirected and unweighted networks. For generalizations we refer to Chaps. A and B.

In order to construct the approximating weight between nodes i and j, one treats the original complex network as a random network and assumes that the coupling strength is proportional to the probability that a node with degree k_i couples to a node with degree k_j. If the degrees are uncorrelated [41, 44, 45], then $\tilde{A}_{ij}$ is equal to the degree k_i times the degree k_j normalized by the total number of all degrees in the network:

$$\tilde{A}_{ij} = k_i \frac{k_j}{\sum_{l=1}^{N} k_l} . \tag{2.5}$$

Apparently, Eq. (2.5) defines a symmetric matrix. It is easy to see that the new adjacency matrix conserves the local degree structure as required in Eq. (2.4). We illustrate the approximation in Fig. 2.1.

Inserting Eq. (2.5) into Eq. (2.1) yields

$$\dot{\phi}_i(t) = \omega_i + \frac{K}{N} \frac{k_i}{\sum_{l=1}^{N} k_l} \sum_{j=1}^{N} k_j \sin\left[\phi_j(t) - \phi_i(t)\right] + \xi_i(t). \tag{2.6}$$

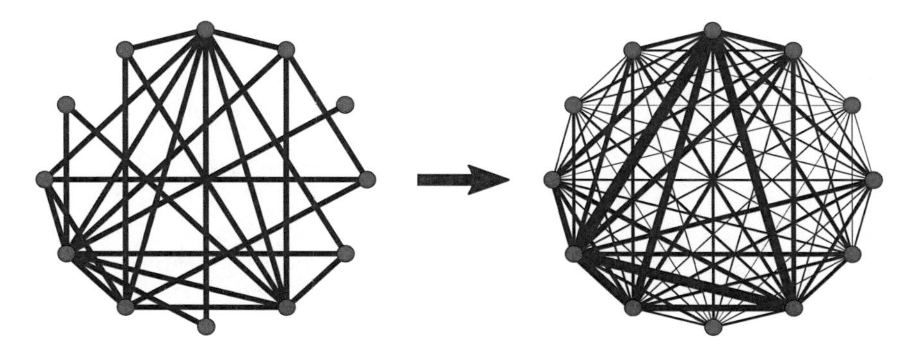

Figure 2.1.: The effect of our approximation on a ring network of eight symmetrically coupled oscillators. On the left-hand side the original complex network is shown and on the right-hand side the approximate network is shown, where the thickness of the edges is chosen approximately proportional to the coupling strength.

Let us take the following definition from Ref. [41] [compare with Eq. (1.3)]:

$$r(t)\mathrm{e}^{i\Theta(t)} := \frac{\sum_{j=1}^{N} k_j \mathrm{e}^{i\phi_j(t)}}{\sum_{j=1}^{N} k_j}. \tag{2.7}$$

The mean-field amplitude $r(t)$ could be used as an order parameter, because for a population of $N \to \infty$ completely asynchronous oscillators, $r(t \to \infty)$ vanishes, while the onset of synchronization is marked by $r(t \to \infty) > 0$ (the perfectly phase synchronized state corresponds to $r(t \to \infty) = 1$)[1].

Multiplying Eq. (2.7) by $\mathrm{e}^{-i\phi_i(t)}$ and considering only the imaginary parts, one can rewrite Eq. (2.6) as an effective one-oscillator description, where the common time-dependent phase $\Theta(t)$ and amplitude $r(t)$ are averaged over all the nodes according to Eq. (2.7):

$$\dot{\phi}_i(t) = \omega_i + r(t)K\frac{k_i}{N}\sin\left[\Theta(t) - \phi_i(t)\right] + \xi_i(t). \tag{2.8}$$

Hence, the network coarse-graining yields a mean-field approximation with the mean field amplitude $r(t)$ and phase $\Theta(t)$ defined in Eq. (2.7). All oscillators are statistically identical and differ only by ω_i and k_i. They are coupled to the mean field with a characteristic strength that is proportional to the individual degree k_i, which can be seen as a property of node i. Such an assignment is similar to a characterization of subpopulations through coupling strengths, cf. Sec. 4.4.

[1] One has to be careful in networks where nodes differ from each other in the way their degrees scale with the system size N. For example, in star graphs with $N-1$ leaves of degree 1 and one hub node with degree $N-1$, the perfectly synchronized state leads to $r = 1/2$, not to $r = 1$. We thank R. Toral for pointing this out.

2.2. The nonlinear Fokker-Planck equation

We start here with a rather general formulation. Instead of the Langevin equations (2.1) the system of N coupled oscillators can be described by a joint probability distribution

$$\mathcal{P}_N\left(\boldsymbol{\phi}, t; \boldsymbol{\phi}^0, t^0; \boldsymbol{\sigma}\right). \tag{2.9}$$

The vector $\boldsymbol{\phi} = (\phi_1, \ldots, \phi_N)$ contains the phases of the N oscillators at time t, while $\boldsymbol{\phi}^0 = (\phi_1^0, \ldots, \phi_N^0)$ consists of their values at the initial time t^0. The time-independent vector $\boldsymbol{\sigma} = (\sigma_1, \ldots, \sigma_N)$ specifies the individual quantities for all oscillators, so each element σ_i is itself a set of properties, cf. Eq. (1.1). Normalization requires

$$\int_0^{2\pi} d\phi_1 \ldots d\phi_N \int_0^{2\pi} d\phi_1^0 \ldots d\phi_N^0 \int d\sigma_1 \ldots d\sigma_N \mathcal{P}_N\left(\boldsymbol{\phi}, t; \boldsymbol{\phi}^0, t^0; \boldsymbol{\sigma}\right) = 1. \tag{2.10}$$

A conditional probability distribution can be obtained as

$$p_N\left(\boldsymbol{\phi}, t | \boldsymbol{\phi}^0, t^0; \boldsymbol{\sigma}\right) = \frac{\mathcal{P}_N\left(\boldsymbol{\phi}, t; \boldsymbol{\phi}^0, t^0; \boldsymbol{\sigma}\right)}{P_N\left(\boldsymbol{\phi}^0, t^0; \boldsymbol{\sigma}\right)}. \tag{2.11}$$

The probability distribution $P_N\left(\boldsymbol{\phi}^0, t^0; \boldsymbol{\sigma}\right)$ gives the probability of the realization of the system concerning all sources of quenched disorder, that is individual quantities $\sigma_1, \ldots, \sigma_N$ and initial conditions $\phi_1^0, \ldots, \phi_N^0$. For given $\sigma_1, \ldots, \sigma_N$ and $\phi_1^0, \ldots, \phi_N^0$, $p_N\left(\boldsymbol{\phi}, t | \boldsymbol{\phi}^0, t^0; \boldsymbol{\sigma}\right)$ denotes the probability to observe oscillators with phases between $\phi_1, \ldots, \phi_N$ and $\phi_1 + d\phi_1, \ldots, \phi_N + d\phi_N$ at time t. In the following we assume independent and identically distributed initial phases ("initial chaos") and independent individual quantities. Then we can factorize

$$P_N\left(\boldsymbol{\phi}^0, t^0; \boldsymbol{\sigma}\right) = \prod_{i=1}^{N} P_{\text{in}}(\phi_i^0)\, P(\sigma_i), \tag{2.12}$$

with some distribution of initial phases, $P_{\text{in}}(\phi_i^0)$, and some distribution for the individual quantities, $P(\sigma_i)$. In the typical mean-field treatment one assumes that a factorization as in Eq. (2.12) can also be applied on the level of the density $\mathcal{P}_N\left(\boldsymbol{\phi}, t; \boldsymbol{\phi}^0, t^0; \boldsymbol{\sigma}\right)$ for arbitrary $t > t_0$. We will come back to this later. For analytical derivations it is useful to introduce reduced probability densities $\mathcal{P}_n$ with index $n = 1, 2, \ldots, N-1$. They are obtained by integrating $\mathcal{P}_N$ over a subset of the phases, the initial conditions and the individual quantities. All oscillators shall be identical in their dynamic behavior with respect to the specific individual quantities. Thus, we can exemplarily take out oscillators with indices $i = 1, \ldots, n$ and integrate over the remaining variables and parameters with $i > n$. This defines the reduced probability densities with the same bounds of integration as in Eq. (2.10):

$$\mathcal{P}_n\left(\phi_1, t; \phi_1^0, t^0; \sigma_1; \ldots; \phi_n, t; \phi_n^0, t^0; \sigma_n\right) = \int \prod_{i=n+1}^{N} \left(d\phi_i d\phi_i^0 d\sigma_i\right) \mathcal{P}_N\left(\boldsymbol{\phi}, t; \boldsymbol{\phi}^0, t^0; \boldsymbol{\sigma}\right). \tag{2.13}$$

The full N-oscillator joint probability density $\mathcal{P}_N\left(\boldsymbol{\phi}, t; \boldsymbol{\phi}^0, t^0; \boldsymbol{\sigma}\right)$ describing the evolution of the population from time t^0 to time $t > t^0$, is governed by a linear Fokker-Planck equation (FPE) [46]:

$$\frac{\partial \mathcal{P}_N}{\partial t} = \sum_{i=1}^{N} \left\{ D \frac{\partial^2 \mathcal{P}_N}{\partial \phi_i^2} - \frac{\partial}{\partial \phi_i} \left[v\left(\boldsymbol{\phi}, t; \boldsymbol{\sigma}\right) \mathcal{P}_N \right] \right\}, \tag{2.14}$$

with some function v specifying the drift. Note that the conditional probability distribution $p_N\left(\boldsymbol{\phi}, t | \boldsymbol{\phi}^0, t^0; \boldsymbol{\sigma}\right)$ [cf. Eq. (2.11)] fulfills the same FPE (2.14), because after inserting it, the distribution $P_N\left(\boldsymbol{\phi}^0, t^0; \boldsymbol{\sigma}\right)$ cancels out. Let us specify the individual properties in the current system. After coarse-graining the network, the oscillators are fully characterized by their natural frequencies and the number of connections they have, i.e. $\sigma_i = (\omega_i, k_i)\,,\ i = 1, \ldots, N$. Hence, the independence between individual quantities is fulfilled, as required in Eq. (2.12). However, on each node, natural frequencies and degrees may depend on each other. The FPE (2.14) for $\mathcal{P}_N\left(\boldsymbol{\phi}, t; \boldsymbol{\phi}^0, t^0; \boldsymbol{\omega}, \boldsymbol{k}\right)$ becomes

$$\frac{\partial \mathcal{P}_N}{\partial t} = D \sum_{i=1}^{N} \frac{\partial^2 \mathcal{P}_N}{\partial \phi_i^2} - \sum_{i=1}^{N} \frac{\partial}{\partial \phi_i} \left\{ \left[\omega_i + \frac{K}{N} \frac{k_i}{\sum_{l=1}^{N} k_l} \sum_{j=1}^{N} k_j \sin\left(\phi_j - \phi_i\right) \right] \mathcal{P}_N \right\}. \tag{2.15}$$

Let us continue with the dynamics for the reduced densities (2.13). To this end, we integrate the FPE (2.15) over the corresponding subset of the variables and the other quantities. Then we obtain a set of coupled differential equations, akin to a Bogoliubov-Born-Green-Kirkwood-Yvon (BBGKY) hierarchy. Truncating this hierarchy at some n leads to a reduced description. Specifically here, we are interested in the one-oscillator probability density $\mathcal{P}_1\left(\phi_1, t; \phi_1^0, t^0; \omega_1, k_1\right)$. Therefore, we integrate the FPE (2.15) over the $N-1$ phases $\phi_2, \ldots, \phi_N$, their initial values $\phi_2^0, \ldots, \phi_N^0$, the natural frequencies $\omega_2, \ldots, \omega_N$ and the degrees[2] $k_2, \ldots, k_N$. This yields

$$\begin{aligned} \frac{\partial \mathcal{P}_1}{\partial t} = &- \omega_1 \frac{\partial \mathcal{P}_1}{\partial \phi_1} - \frac{K}{N} \frac{k_1 (N-1)}{\sum_l k_l} \frac{\partial}{\partial \phi_1} \int \mathrm{d}\phi_2 \int \mathrm{d}\phi_2^0 \int d\omega_2 \sum\nolimits_{k_2} \sin\left(\phi_2 - \phi_1\right) \\ &\times k_2 \mathcal{P}_2\left(\phi_1, \phi_2, t; \omega_1, \omega_2, k_1, k_2\right) + D \frac{\partial^2 \mathcal{P}_1}{\partial \phi_1}. \end{aligned} \tag{2.16}$$

Here we use the statistical identity of all the oscillators as mentioned when writing down Eq. (2.13). Note that $\mathcal{P}_1$ is related hierarchically to $\mathcal{P}_2$ being the two-oscillator distribution. The essence of a mean-field theory lies in neglecting dynamical correlations, which here means to identify

$$\mathcal{P}_2\left(\phi_1, \phi_2, t; \phi_1^0, \phi_2^0, t^0; \omega_1, k_1, \omega_2, k_2\right) \equiv \mathcal{P}_1\left(\phi_1, t; \phi_1^0, t^0; \omega_1, k_1\right) \mathcal{P}_1\left(\phi_2, t; \phi_2^0, t^0; \omega_2, k_2\right). \tag{2.17}$$

Such a factorization goes beyond Eq. (2.12) due to the dependence on time t. It corresponds essentially to the lowest-order truncation of the BBGKY hierarchy. Equation (2.16) becomes closed but nonlinear in $\mathcal{P}_1$. Remarkably, in the thermodynamic limit of

[2] In case of infinitely many different degrees one could approximate the sum over degrees by an integral.

infinitely many oscillators, $N \to \infty$, such a truncation can be justified in a rigorous way for various systems (see Refs. [47–49] and references therein for Kuramoto-type systems and Refs. [50, 51] for recent and more general overviews along with new results). Here it suffices to say that the argument goes back to Boltzmann's "Stosszahlansatz", which was later rigorously formalized by Kac under the concept of "propagation of molecular chaos" [52]. In summary, Eq. (2.17) can be considered to be exact in the thermodynamic limit. Henceforth we can neglect the indices at ϕ, ω and k, as the underlying assumption in the mean-field approach is that oscillators with the same natural frequency and degree are statistically identical. We continue with the conditional form of the one-oscillator probability density, $p_1(\phi, t|\phi^0, t^0; \omega, k) = \mathcal{P}_1(\phi, t; \phi^0, t^0; \omega, k) / [P_{\text{in}}(\phi^0) P(\omega, k)]$ [see Eq. (2.11)]. For every given pair (ω, k), the expression $p_1(\phi, t|\phi^0, t^0; \omega, k)\, d\phi$ denotes the fraction of oscillators, which start with the phase ϕ^0 at time t^0 and then have a phase value between ϕ and $\phi + d\phi$ at time t. The normalization (2.10) now reads $1 = \int_0^{2\pi} d\phi\, p_1(\phi, t|\phi^0, t^0; \omega, k)\ \forall\, t, \phi^0, \omega, k$. Note that the ratio $\sum_l k_l/(N-1)$ becomes the average degree for large N. Finally, we can write down a McKean-Vlasov or nonlinear Fokker-Planck equation [53] for the dynamical evolution of p_1:

$$\frac{\partial p_1}{\partial t} = -\frac{\partial}{\partial \phi}\left[v_{\omega,k}(\phi, t) p_1\right] + D\frac{\partial^2 p_1}{\partial \phi^2}. \tag{2.18}$$

The drift $v_{\omega,k}(\phi, t)$, i.e. the mean increment of the phase per unit time, is given by

$$v_{\omega,k}(\phi, t) = \omega + r(t) K \frac{k}{N} \sin\left[\Theta(t) - \phi(t)\right]. \tag{2.19}$$

Equation (2.18) is called nonlinear, because the drift depends on the density p_1 via the mean-field amplitude $r(t)$ and phase $\Theta(t)$,

$$r(t) e^{i\Theta(t)} = \frac{\langle r_{\omega',k'}(t)\ k'\ e^{i\Theta_{\omega',k'}(t)}\rangle_{\omega',k'}}{\langle k' \rangle}. \tag{2.20}$$

The double average $\langle \ldots \rangle_{\omega',k'} \equiv \int d\omega' \sum_{k'} \ldots P(\omega', k')$ connects in a superposed manner the global with the following local mean-field variables,

$$r_{\omega,k}(t) e^{i\Theta_{\omega,k}(t)} = \int_0^{2\pi} d\phi \int_0^{2\pi} d\phi^0 e^{i\phi} p_1\left(\phi, t|\phi^0, t^0; \omega, k\right) P_{\text{in}}(\phi^0). \tag{2.21}$$

The set of equations (2.18)-(2.21) has to be solved with the initial condition for the transition probability density

$$p_1\left(\phi, t^0|\phi^0, t^0; \omega, K\right) = \delta\left(\phi - \phi^0\right), \forall\, \omega, K. \tag{2.22}$$

So far we have not dropped the dependence of the conditional probability density on the initial state. Since the FPE is nonlinear, the temporal evolution of the system might sensitively depend on the distribution of initial phases, e.g. in case of bistability. In the following analytical treatment we do not investigate this circumstance further. Bistability, which is not an issue in this chapter, will be analyzed in terms of lower-dimensional systems of coupled ODE's for the mean-field variables. Instead, we formulate the problem in terms of a nonlinear FPE for the marginal density of the phase ϕ at time t,

$$\rho\left(\phi, t|\omega, k\right) \equiv \int_0^{2\pi} d\phi^0 p_1\left(\phi, t|\phi^0, t^0; \omega, k\right) P_{\text{in}}(\phi^0). \tag{2.23}$$

Specifically, this distribution ρ replaces the conditional probability density in the nonlinear FPE (2.18) and in Eq. (2.21) via integration over the initial phases. As a result, we are left with an initial value problem that has to be solved under the initial condition

$$\rho\left(\phi, t^0|\omega, k\right) = P_{\text{in}}(\phi^0)\,,\ \forall\ \omega, k\,. \tag{2.24}$$

We would like to emphasize that the nonlinear FPE (2.18) comprises a large system of coupled partial differential equations. The oscillators with the same natural frequency and the same degree can be interpreted as one species. If there are infinitely many different natural frequencies or degrees, the problem is infinitely dimensional. Every species obeys the FPE with the corresponding ω and k. They contribute with subfields given by Eq. (2.21) in accordance with their emergence to the mean field, Eq. (2.20), which is given by averaging the subfields over the joint probability distribution $P(\omega, k)$.

2.3. The critical coupling strength

We are interested here in the onset of synchronization. This is when in a completely incoherent population of oscillators a macroscopic fraction, i.e. a number of order $O(N)$, transitions to collective oscillations. The critical condition is met when the incoherent solution becomes unstable. Therefore, we will perform a linear stability analysis of the incoherent solution. The following analysis is an extension of the derivations in the work of Strogatz and Mirollo [54] by allowing complex networks in a coarse-grained way, cf. 2.1. Recall first the normalization condition

$$\int_0^{2\pi} \rho(\phi, t|\omega, k) d\phi = 1\ \ \forall\ \omega, k, t\ . \tag{2.25}$$

In the completely incoherent state any phase value is equally probable, hence the phase distribution in the stationary regime (denoted by index 0) is given by

$$\rho_0^{\text{incoh}}(\phi|\omega, k) = \frac{1}{2\pi}\ \ \forall\ \omega, k, \tag{2.26}$$

with vanishing order parameters $r_{\omega,k} = 0$ and undefined mean phases $\Theta_{\omega,k}$, cf. Eq. (2.21). It is easy to see that ρ_0^{incoh} is at least one of the possible asymptotic solutions of the nonlinear FPE (2.18).

Consider the evolution of a small perturbation of the incoherent state:

$$\rho(\phi, t|\omega, k) = \frac{1}{2\pi} + \epsilon\delta\rho(\phi, t|\omega, k),\ \ \epsilon \ll 1. \tag{2.27}$$

The normalization condition (2.25) implies

$$\epsilon \int_0^{2\pi} \delta\rho(\phi, t|\omega, k) d\phi = 0 \;\forall\; \omega, k, t \;, \tag{2.28}$$

so $\delta\rho(\phi, t|\omega, k)$ is 2π-periodic in ϕ. Now we substitute Eq. (2.27) into the FPE (2.18),

$$\epsilon \frac{\partial \delta\rho}{\partial t} = -\frac{\partial}{\partial \phi}\left[\left(\frac{1}{2\pi} + \epsilon\delta\rho\right) v_{\omega,k}\right] + \epsilon D \frac{\partial^2 \delta\rho}{\partial \phi^2} \;. \tag{2.29}$$

The amplitude becomes $r(t) = \epsilon\delta r(t)$, where $\delta r(t)$ is given by substituting ρ by $\delta\rho$ in Eq. (2.20). The FPE (2.18) linearized in the lowest order in ϵ then reads

$$\frac{\partial \delta\rho}{\partial t} = -\omega \frac{\partial \delta\rho}{\partial \phi} + \frac{Kk}{2\pi N} \delta r \cos(\Theta - \phi) + D \frac{\partial^2 \delta\rho}{\partial \phi^2} \;. \tag{2.30}$$

Since $\delta\rho$ is a real number and 2π-periodic in ϕ [cf. Eq. (2.28)], it may be expanded as

$$\delta\rho(\phi, t|\omega, k) = \frac{1}{2\pi} \sum_{l=1}^{\infty} \left\{ c_l(t|\omega, k) \mathrm{e}^{il\phi} + c_l^*(t|\omega, k) \mathrm{e}^{-il\phi} \right\} \;. \tag{2.31}$$

Due to Eq. (2.28), $c_0(t|\omega, k)$ vanishes. The first nonvanishing coefficients, which constitute the so-called fundamental mode, determine the deviations of the mean field in first order of small ϵ:

$$\delta r \mathrm{e}^{i\Theta} = \frac{1}{\langle k \rangle} \int_{-\infty}^{+\infty} d\omega' \sum_{k'=k_{\min}}^{k_{\max}} c_1^*(t|\omega', k') k' P(\omega', k') \;. \tag{2.32}$$

Here $k_{\min}$ and $k_{\max}$ are the minimum and maximum degree in the network, respectively. In order that the incoherent state becomes unstable and a synchronization process starts, $c_1(t|\omega, k)$ has to grow, so that as a consequence the order parameter $r(t)$ grows as well.

Multiplication of the last equation by $\mathrm{e}^{-i\phi}$ and considering only the real parts, yields

$$\delta r \cos(\Theta - \phi) = \frac{1}{2\langle k \rangle} \left(\int_{-\infty}^{+\infty} d\omega' \sum_{k'=k_{\min}}^{k_{\max}} c_1(t|\omega', k') k' P(\omega', k') \right) \mathrm{e}^{i\phi} + \mathrm{c.\,c.} \;. \tag{2.33}$$

Now we insert Eqs. (2.31)-(2.33) into Eq. (2.30) and consider only the coefficients with $\mathrm{e}^{i\phi}$. Afterwards, one finds the evolution equation for the fundamental mode $c_1(t|\omega, k)$:

$$\frac{\partial c_1}{\partial t} = -(D + i\omega) c_1 + \frac{Kk}{2\langle k \rangle N} \int_{-\infty}^{+\infty} d\omega' \sum_{k'=k_{\min}}^{k_{\max}} c_1(t|\omega', k') k' P(\omega', k') \;. \tag{2.34}$$

Since this is a partial differential equation without mixed derivatives, we make a separation ansatz:

$$c_1(t|\omega, k) \equiv b(\omega, k)\; \mathrm{e}^{\lambda t}, \;\; \lambda \in \mathbb{C}. \tag{2.35}$$

Thus, we obtain

$$\lambda b = -(D + i\omega) b + \frac{Kk}{2\langle k \rangle N} \int_{-\infty}^{+\infty} d\omega' \sum_{k'=k_{\min}}^{k_{\max}} b(\omega', k') k' P(\omega', k') \;. \tag{2.36}$$

The right-hand side is a linear transformation and it can be shown that we only have to calculate its point spectrum in order to obtain the critical coupling strength [54].

The second term on the right-hand side of Eq. (2.36) equals the degree k times a constant which we call B in the following, so that $b(\omega, k)$ is given by

$$b(\omega, k) = \frac{Bk}{\lambda + D + i\omega} \, . \tag{2.37}$$

With this, we can set up the following self-consistent equation:

$$B = \frac{K}{2\langle k\rangle N} \int_{-\infty}^{+\infty} d\omega' \sum_{k'=k_{\min}}^{k_{\max}} \frac{Bk'^2 P(\omega', k')}{\lambda + D + i\omega'} \, . \tag{2.38}$$

This relation holds, if all the oscillators are statistically identical, see section 2.2. In Eq. (2.38) the first integral exists for all $\mathrm{Re}(\lambda) > -D$ [55]. Setting $B = 0$ would lead to the trivial solution $c_1(t|\omega, k) = 0$. Therefore we can divide Eq. (2.38) by B:

$$1 = \frac{K}{2\langle k\rangle N} \int_{-\infty}^{+\infty} d\omega' \sum_{k'=k_{\min}}^{k_{\max}} \frac{k'^2 P(\omega', k')}{\lambda + D + i\omega'} \, . \tag{2.39}$$

Analogous to the proof in [56], one can show that there exists only one solution for λ and that it has to be a real number. The proof makes use of the assumption that $P(\omega, k)$ is symmetric and unimodal with respect to ω. Due to the rotational symmetry in the model the most probable natural frequency can be located at $\omega = 0$ without loss of generality. In summary, the eigenvalue λ obeys

$$1 = \frac{K}{2\langle k\rangle N} \int_{-\infty}^{+\infty} d\omega' \sum_{k'=k_{\min}}^{k_{\max}} \frac{(\lambda + D)k'^2 P(\omega', k')}{(\lambda + D)^2 + \omega'^2} \, . \tag{2.40}$$

Note that this equation can only be fulfilled if $\lambda > -D$, because otherwise the right-hand side would be nonpositive. Thus, the eigenvalue λ is strictly positive and the incoherent solution cannot be linearly stable, if the noise intensity D vanishes.

At the critical condition $\lambda = \lambda_c = 0$, the incoherent solution becomes unstable: if $\lambda > 0$ the perturbation $\delta\rho$ of the incoherent solution grows exponentially with time $\sim \mathrm{e}^{\lambda t}$ and so does the order parameter $r(t)$ (cf. Eq. (2.32)). Hence, the critical coupling strength K_c for the onset of synchronization is given by

$$K_c = 2N\langle k\rangle \left[\int_{-\infty}^{+\infty} d\omega' \sum_{k'=k_{\min}}^{k_{\max}} \frac{Dk'^2 P(\omega', k')}{D^2 + \omega'^2} \right]^{-1} \, . \tag{2.41}$$

We emphasize that this equation is not valid in the noise-free case, where one has to take the limit $\lambda \to 0^+$ in Eq. (2.40) with $D = 0$, by which previous results can be reproduced [41, 57] (compare Tab. 2.1).

In Sec. 2.6 we investigate an example where the degrees $\boldsymbol{k}$ and the frequencies $\boldsymbol{\omega}$ are not independent. Here we continue with two separated distributions $g(\omega)$ and $P(k)$. It is interesting to see that the critical coupling strength K_c can then be written as a product of two functionals. The first one maps the degree distribution $P(k)$ to a real number via the first two moments,

$$f_{\text{top}}[P] := N\frac{\langle k\rangle}{\langle k^2\rangle}. \tag{2.42}$$

We call this one the "topology functional". The second one maps the frequency distribution $g(\omega)$ to a real number via an integral that depends on the noise intensity D,

$$f_{\text{div}}(D)[g] := 2\left[\int_{-\infty}^{+\infty} d\omega' \frac{Dg(\omega')}{D^2+\omega'^2}\right]^{-1}. \tag{2.43}$$

We call this one the "diversity functional". In short, we have

$$K_c = f_{\text{top}}[P]f_{\text{div}}(D)[g]. \tag{2.44}$$

2.4. Application to dense small-world networks

We would like to test the analytical expression for the critical coupling strength (2.42)-(2.44). To this end, we consider first locally coupled ring-like networks, where random shortcuts to other nodes are added with a certain probability. In particular, we introduce a dense small-world network model. We apply the definition [58] that the total number of edges in dense networks of N nodes scales with N^2. Our dense small-world networks are constructed as follows. Each oscillator is coupled to its z next neighbors in both directions of the ring network and z is given by

$$z = \left\lfloor \frac{\alpha}{2}(N-1)\right\rfloor, \quad 0 \leq \alpha \leq 1. \tag{2.45}$$

The system-size independent parameter α gives the fraction of all nodes that are coupled through next neighbor connections; $\alpha = 0$ yields a network where all nodes are uncoupled, whereas in case of $\alpha = 1$ the network is fully connected. The floor function $\lfloor . \rfloor$ gives the greatest integer less than or equal to its argument, by which the continuous parameter α is mapped to the discrete parameter z. So far we obtain a $2z$-regular network. In such a non-random network, z stands for the coupling radius. With self-coupling all nodes have in total $2z+1$ connections and the degree distribution is just the Kronecker delta symbol,

$$P_{\text{local}}(k) = \delta_{k,2z+1}. \tag{2.46}$$

In our final simulations we choose $\alpha = 0.05$. Then 2.5 percent of possible edges are local regular connections. Besides these $(2z+1)N$ non-random local connections we introduce random shortcuts: each oscillator is coupled to the remaining other $N-2z-1$ ones with a certain probability p.

The shortest path length between two nodes is the minimal number of links connecting them. The topology of a network is often characterized by two quantities: first, the characteristic path length (sometimes called the average shortest path length)

$$L = \frac{2}{N(N-1)}\sum_{i=1}^{N-1}\sum_{j=i+1}^{N} \text{spl}_{ij}, \tag{2.47}$$

where spl_{ij} is the shortest path length between vertices i and j [59–61] and second, the clustering coefficient

$$C = \frac{2}{N} \sum_{i=1}^{N} \frac{m_i}{k_i\,(k_i - 1)}, \tag{2.48}$$

where k_i is the degree and m_i the number of connections within the set of nodes which are coupled to node i. By definition, C lies between zero and unity, whereas L is greater than unity, because the shortest path length between two different nodes is always greater or equal to one.

If the characteristic path length is small, the network is said to have the small-world property. Many real networks have this property (Ref. [59], Table 1). One can observe that in real networks the small-world property is often associated with large values of the clustering coefficient (Ref. [61], Table 2.1). On account of this, Watts and Strogatz [62] proposed to define small-world networks as those networks having a small characteristic path length, but a high clustering coefficient. The valuations "small" or "high" usually arise from the comparison with random networks that have the same number of vertices and the same average degree. In this way a small-world network is characterized by

$$L \gtrsim L_{\mathrm{random}} \text{ and } C \gg C_{\mathrm{random}}. \tag{2.49}$$

In Fig. 2.2 we provide numerical evidence that densely connected networks obtained with the aforementioned procedure [see Eqs. (2.45), (2.46) and explanations there], indeed can fall into the class of small-world networks. In order to do so, we have to compare these networks with random networks. Importantly, the random network used for the comparison should not have any isolated nodes. Otherwise path length and clustering coefficient are not (well) defined. We take as an almost random network, similar to Watts and Strogatz [62], a network generated with our model, but with a minimum amount of coupled next neighbors, i.e. we choose $z = 1$ or an equivalent α [cf. Eq. (2.45)]. The comparison makes only sense, if the average degrees $\langle k \rangle$ are the same in both networks, which requires to take the following shortcut probability p' for the random network:

$$p' = \frac{2(z-1)}{N-3} + \frac{N-2z-1}{N-3} p\,. \tag{2.50}$$

p is the shortcut probability used for the dense small-world network.

Having in mind the defining properties of a small-world network (2.49), we observe small-world networks for values of p around 10^{-3}. For smaller values of α, i.e., fewer non-random local connections, the defining properties are better fulfilled. If α is too large the network under consideration cannot be called small-world network anymore.

If z equals zero the model generates Erdős-Rényi random networks. A similar small-world network model was recently investigated in [64]. The case $p = 1$ leads to all-to-all connectivity differently than in the popular sparse small-world network models [62, 65]. The procedure of adding shortcuts can be seen as a Bernoulli experiment with probability p of success [59–61]. Therefore, the degree distribution $P(k)$ is given by the following binomial distribution:

$$P(k) = \binom{N-2z-1}{k-2z-1} p^{k-2z-1} \left[1-p\right]^{N-k}, \quad k > 2z. \tag{2.51}$$

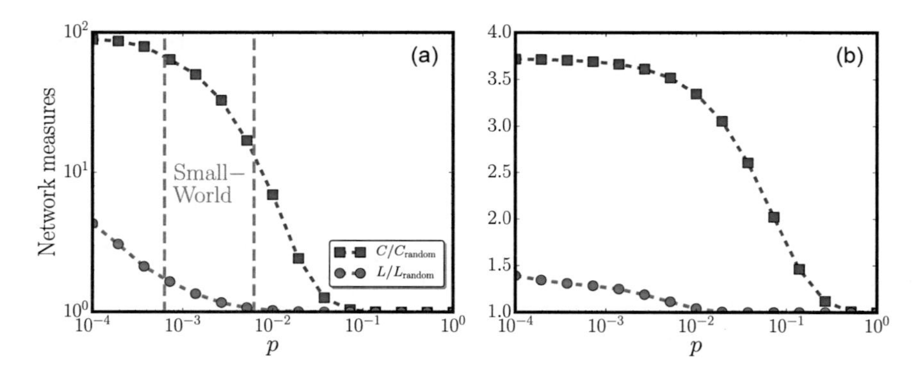

Figure 2.2.: Numerically calculated average path lengths L and clustering coefficients C as a function of shortcut probability p for a network of 501 nodes. The radius of coupled next neighbors [compare Eq. (2.45)] equals $z = 2$ in panel (a) and $z = 50$ in panel (b). Dashed lines indicate the area, in which our model generates a small-world network. Data from [63].

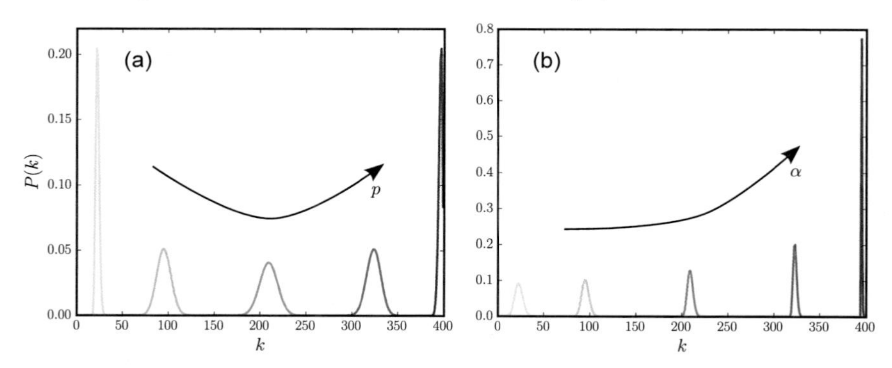

Figure 2.3.: Degree distribution $P(k)$ of dense small-world networks consisting of 400 nodes for different shortcut probabilities p and fractions of regular connections α. (a) $\alpha = 0.05$ and $p = \{0.01, 0.2, 0.5, 0.8, 0.99\}$. (b) $p = 0.05$ and $\alpha = \{0.01, 0.2, 0.5, 0.8, 0.99\}$. For the sake of clarity solid lines are chosen. Data from [63].

Note that the minimum degree $k = 2z + 1$ corresponds to Eq. (2.46). The first moment equals

$$\langle k \rangle = 2z + 1 + (N - 2z - 1)p \approx (N - 1)(\alpha + p - \alpha p) + 1. \tag{2.52}$$

In the last step we have used Eq. (2.45) by neglecting the floor function as an approximation. For large N the latter expression approaches $\langle k \rangle \approx N(\alpha + p - \alpha p)$.

After substitutions one finds the second moment

$$\langle k^2 \rangle = \langle k \rangle^2 + (\langle k \rangle - 2z - 1)(1 - p) , \tag{2.53}$$

where the second item equals the variance of k in the binomial distribution due to the randomness of the number of shortcuts. It scales only linearly with the system size, whereas the first item grows with N^2. Thus, for large systems the difference between the nonrandom number of local connections and the number of shortcuts disappears, because the variability of the shortcuts does not count for large N. The second moment approaches $\langle k \rangle^2$ and it becomes symmetric in α and p. Hence, the topology function [cf. Eq. (2.42)] for large N results in

$$f_{\text{top}}(p, \alpha)|_{N \gg 1} \approx \frac{N}{\langle k \rangle} \approx \frac{1}{\alpha + p - \alpha p} . \tag{2.54}$$

We obtain as limiting cases (compare Fig. 2.4)

$$f_{\text{top}}(p, \alpha)|_{N \gg 1} \approx \begin{cases} \frac{1}{\alpha} + \frac{\alpha - 1}{\alpha^2} p + O(p^2), & p \ll 1, \\ 2 - \alpha + (\alpha - 1)p + O((1-p)^2), & p \lesssim 1. \end{cases} \tag{2.55}$$

Due to the symmetry, we obtain qualitatively the same for $\alpha \ll 1$ and $\alpha \lesssim 1$, but with p and α being interchanged. As expected, the topology function tends to unity for $p \to 1$ or $\alpha \to 1$, because in both cases the network becomes fully connected.

In Fig. 2.4 the dependence of the topology function on α and p is depicted. Discrepancies between numerical calculations and theory in the panel (b) are due to the fact that the number of coupled next neighbors is, of course, an integer. Rounding off in Eq. (2.45) leads to noticeable steps in $f_{\text{top}}(p, \alpha)$ for smaller N.

We emphasize that with the help of Eqs. (2.52) and (2.53), one can find the exact expression for the topology function for arbitrary system sizes. Since it is a rather lengthy expression, we skip it in the text but we will use it in Fig. 2.4. Moreover, as can be seen in Fig. 2.4, a system size N of the order $O(100)$ is sufficient to obtain a dynamical behavior that is comparable with the thermodynamic limit. The critical coupling strength K_c has been derived in the thermodynamic limit [see Eq. (2.44)]. For this reason it is only consistent to calculate the topology function in the limit $N \to \infty$ as well.

In our simulations, the stochastic differential equations (2.1) are integrated up to $t = 600$ with time step $h = 0.05$ by using the Heun scheme [66, 67]. The periods of the oscillators are $T \sim O(10)$, so our integrations cover $O(10)$ periods. Moreover, in order to calculate statistical equilibria, we discard the data up to $t = 200$, by which transient effects are safely avoided. The statistical equilibria are further calculated as averages over at least 100 different network realizations. We emphasize that all the different network configurations do not differ only in the configuration of the connections, but the oscillators on the network differ as well: all the natural frequencies and the initial values of the phases change from one configuration to another one.

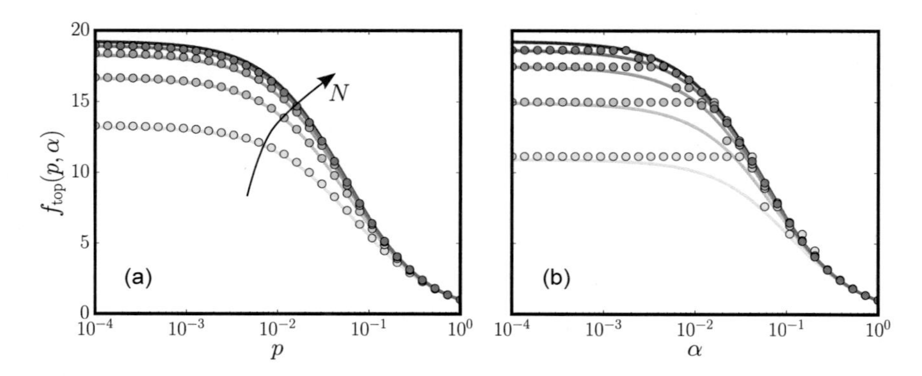

Figure 2.4.: For system sizes $N = \{40, 117, 350, 1120\}$, it is shown the dependency of the topology function on (a) the shortcut probability p with fixed coupling radius $\alpha = 0.052$, and (b) on coupling radius α with fixed shortcut probability $p = 0.052$. All solid lines represent theoretical results. The thick line corresponds to the thermodynamic limit and slight lines are calculated according to Eqs. (2.52) and (2.53) for smaller systems. Markers show results of numerical calculations for indicated system sizes. Data from [63].

Motivated by the works presented in [68–71], we calculate the order parameter averaged over time and network realizations and use the following scaling form:

$$\langle r(t)\rangle_{t,\mathcal{E}} = \left\langle \left| \frac{1}{N}\sum_{j=1}^{N} \mathrm{e}^{i\phi_j} \right| \right\rangle_{t,\mathcal{E}} = N^{-\frac{\beta}{\nu}} F\left[(K-K_c)N^{\frac{1}{\nu}}\right] \,, \tag{2.56}$$

where $F[.]$ is some scaling function. From Refs. [68–71] we know that β and ν should assume the values $\beta \approx \frac{1}{2}$ and $\nu \approx 2$.

Eq. (2.56) proposes a finite-size scaling analysis to calculate the critical coupling strength, because at $K = K_c$ the function $F[.]$ is independent of the system size N. By plotting $\langle r(t)\rangle_{t,\mathcal{E}} N^{0.25}$ as a function of K for various network sizes N, we can measure the critical coupling strength K_c as a well-defined intersection point, see Fig. 2.5.

In order to compare the theory, expressed by the critical coupling strength (2.44), with simulations on the proposed dense small-world networks, we use as topology functional the approximation for large systems [cf. Eq. (2.54)].

In the following we further consider a Gaussian frequency distribution $g_{\text{gauss}}(\omega)$ with vanishing mean and some standard deviation σ. Then for the diversity functional (2.43) the expression

$$f_{\text{div}}(D)[g_{\text{gauss}}] = 2\sqrt{\frac{2}{\pi}}\sigma \left[1 - \Phi\left(\frac{D}{\sqrt{2}\sigma}\right)\right]^{-1} \mathrm{e}^{-\frac{1}{2}\frac{D^2}{\sigma^2}} \tag{2.57}$$

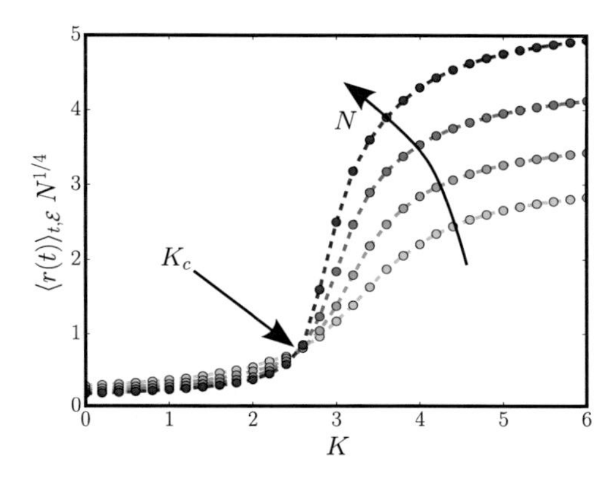

Figure 2.5.: Numerical determination of critical coupling strength by plotting the order parameter $\langle r(t)\rangle_{t,\mathcal{E}}$ times $N^{0.25}$ for different system sizes in dependence on the coupling strength. According to Eq. (2.56) the intersection of the curves indicates the critical coupling strength. Parameter values here: $\alpha = 0.05$, $p = 0.1$, $D = 0.05$, $\sigma = 0.02$. Data from [63].

follows; $\Phi(.)$ is the error function. The formula is equivalent to the mean field expression derived in [54] by means of an eigenvalue analysis.

As can be seen in Fig. 2.6, we obtain a satisfying agreement for the critical coupling strength K_c. Especially for $p > 0.1$, $\alpha > 0.1$ or $\sigma < 0.5$ and for the dependency on the noise intensity D in general, we obtain almost a perfect agreement between theory and simulations. For smaller values of p or α there is a small discrepancy, but the shape of the curves can be well reproduced.

As expected, both weaker connectivity and larger diversity impede synchronization. Compare Eq. (2.55) and Tab. 2.1 for a summary of the limiting cases. In particular, for $p \to 0$ or $\alpha \to 0$ our results suggest a saturation at the values $K_c = f_{\text{div}}(D)[g]/\alpha$ or $K_c = f_{\text{div}}(D)[g]/p$, respectively.

Apparently, the weighted mean-field approximation tends to overestimate the critical coupling strength, which is counterintuitive, because the approximation corresponds to a weighted fully connected network and all-to-all connectivity should reduce the critical coupling strength. To see this, compare $p = 1$ or $\alpha = 1$ in Figs. 2.6(a,b), because both choices stand for all-to-all connectivity. So the coupling weights defined in Eq. (2.5) are able to mimic the original complexity in an overstated manner.

The disagreement for high standard deviations σ of the Gaussian frequency distribution seems to be analogous to the disagreement in the dependency on the topology, because in our derivation of the critical coupling strength, both the frequency distribution $g(\omega)$ and the degree distribution $P(k)$ are involved in averaging.

In what follows, we consider the diversity functional for different frequency distributions:

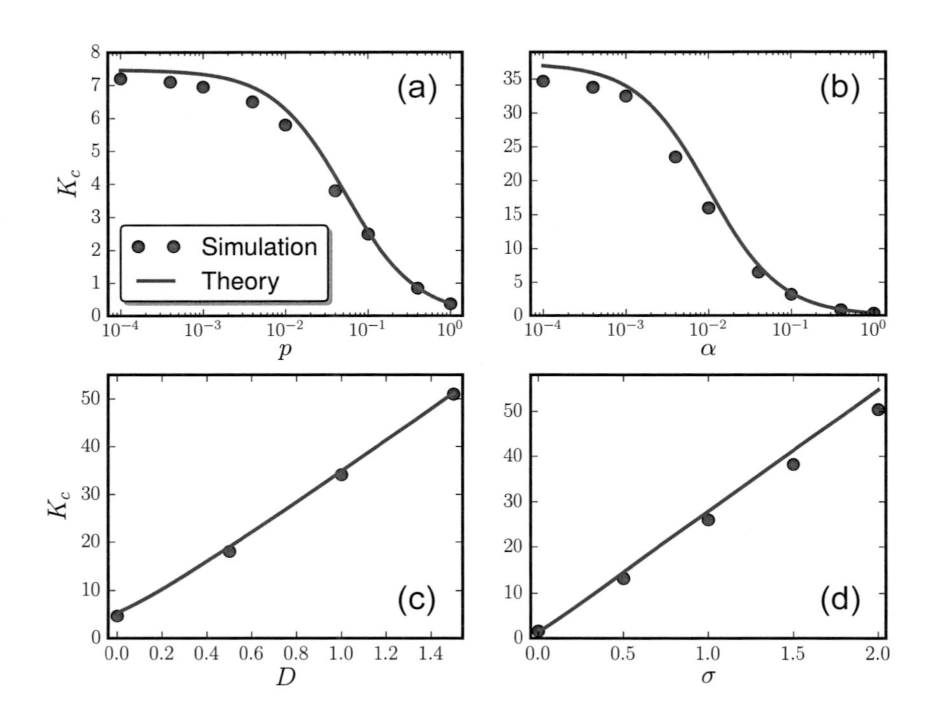

Figure 2.6.: Summary on how the critical coupling K_c depends on the various parameters. In panel (a) the shortcut probability p is varied with fixed coupling radius $\alpha = 0.05$, while in panel (b) the coupling radius α is varied with fixed shortcut probability $p = 0.01$. In both panels the remaining parameters are noise intensity $D = 0.05$ and frequency standard deviation $\sigma = 0.2$. In panel (c) the noise intensity D is varied with fixed $\sigma = 0.2$, and in panel (d) σ is varied with fixed $D = 0.05$. Remaining parameters there: $\alpha = 0.05$, $p = 0.01$. Data from [63].

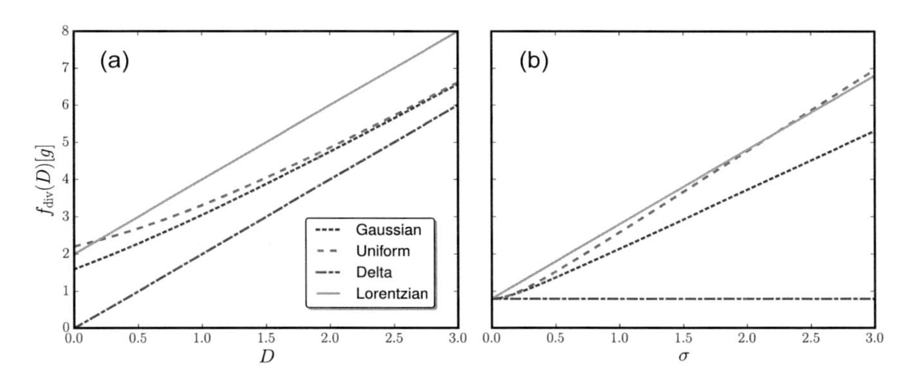

Figure 2.7.: The diversity functional for different frequency distributions as a function of noise intensity D, and standard deviation / scale parameter σ. Data from [63].

uniform distribution $g_{\text{uni}}(\omega)$, identical oscillators $g_{\text{ident}}(\omega)$ and Lorentzian distribution $g_{\text{lorentz}}(\omega)$. We obtain the following expressions for the diversity functionals [cf. Eq. (2.43)]:

$$\begin{aligned} f_{\text{div}}(D)[g_{\text{uni}}] &= 2\sqrt{3}\sigma\left[\arctan\left(\frac{\sqrt{3}\sigma}{D}\right)\right]^{-1}, \\ f_{\text{div}}(D)[g_{\text{ident}}] &= 2D, \\ f_{\text{div}}(D)[g_{\text{lorentz}}] &= 2(D+\gamma). \end{aligned} \tag{2.58}$$

σ and γ are the standard deviation and the scale parameter, respectively, while D gives the noise intensity. In Fig. 2.7 the dependencies on D, σ or γ are depicted. We observe a greater difference in the dependency on diversity than on noise. In particular, for large values of D, all the different diversity functionals show the same linear dependency on D [cf. Fig. 2.7(a)]. In case of Gaussian, uniformly and identically distributed natural frequencies ω, it can be shown that the diversity functional even approaches the same line for $D \to \infty$.

However, the above observation is not surprising, because the noise acts on the natural frequencies. So if the noise intensity D is very high, it makes almost no difference how the natural frequencies are distributed, because the diversity of the oscillators mainly comes from the random fluctuations induced by the noise.

Instead, for $D =$ const. and increasing diversity parameter σ or γ, the different nature of the various frequency distributions manifests more and more in a different synchronizability [cf. Fig. 2.7(b)]. Especially the comparison between the uniform and the Lorentzian distribution for the same σ and γ values is interesting. In terms of synchronizability we observe that for lower diversity the uniform frequency distribution is more favorable than the Lorentzian distribution, while for higher diversity the Lorentzian distribution is more favorable.

Of course, vanishing diversity, i.e., $\sigma \to 0$ or $\gamma \to 0$, results in identical oscillators, so that the diversity functional approaches the same value for every frequency distribution. In Tab. 2.1 we summarize these results.

LIMITS	$D \ll 1$	$\sigma \ll 1$	$D \gg 1$	$\sigma \gg 1$
g_{gauss}	$2\sqrt{\frac{2}{\pi}}\sigma + \frac{4}{\pi}D + O(D^2)$	$2D + \frac{2}{D}\sigma^2 + O(\sigma^4)$	$2D + O\left(\frac{1}{D}\right)$	$2\sqrt{\frac{2}{\pi}}\sigma + O(1)$
g_{uni}	$\frac{4\sqrt{3}}{\pi}\sigma + \frac{8}{\pi^2}D + O(D^2)$	$2D + \frac{2}{D}\sigma^2 + O(\sigma^4)$	$2D + O(\frac{1}{D})$	$\frac{4\sqrt{3}}{\pi}\sigma + O(1)$
g_{ident}	$2D$	$2D$	$2D$	$2D$
g_{lorentz}	$2\sigma + 2D$	$2D + 2\sigma$	$2D + O(1)$	$2\sigma + O(1)$

Table 2.1.: The limiting cases of the diversity functional for different frequency distributions. Here $\sigma = \gamma$.

2.5. Interpolating between sparse and dense networks

Since the mean-field description corresponds to all-to-all connectivity, we expect that the approximation loses validity in sparser networks. In order to effectively analyze the limitations, we consider separately Erdős-Rényi-like random networks and regular networks. The first are constructed by assigning an edge probability of

$$p_e = pN^{q-1}, \; 0 \le p, q \le 1 \tag{2.59}$$

for any two of the N nodes in the network. The only additional requirement is that besides the edge probability, each node connects a priori to another randomly chosen one. In this way we guarantee that there are no isolated nodes, which are not of interest here, because they are not able to take part in the synchronization process and only reduce the effective system size.

In the regular networks each node is coupled to the z next neighbors in both directions of the ring network and z is given by [compare Eq. (2.45)]

$$z = \left\lfloor \frac{1}{2}\left[3 + \alpha(N-4)^q\right] \right\rfloor, \; 0 \le \alpha, q \le 1. \tag{2.60}$$

As for the random networks, we require that there are no isolated nodes, which is already implemented in the above choice ($z \ge 1$).

In both cases q is a denseness parameter, for $q = 1$ leads to dense and $q = 0$ to sparse networks [58].

We repeat the former analysis, see Sec. 2.3, with adjusting the normalization of the coupling, i.e., we substitute $N \to N^q$ in front of the coupling term [cf. Eq. (2.1)]. Consequently, the topology functional becomes

$$f_{\text{top}}[P] = N^q \frac{\langle k \rangle}{\langle k^2 \rangle}. \tag{2.61}$$

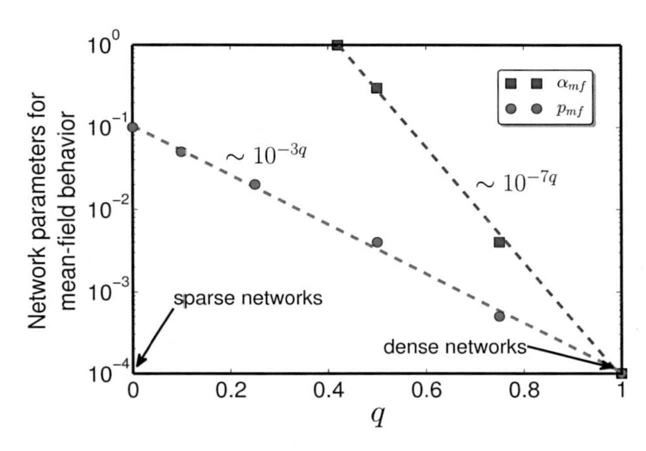

Figure 2.8.: Smallest p (random network) and α (regular network) values for which a synchronization transition with the standard mean-field exponents can be observed, are shown as functions of the denseness parameter q. We denote these values by p_{mf} and α_{mf}. Dashed lines depict the scaling behaviors. Data from [63].

In case of the regular networks it appears necessary to increase the system size for sparser networks in order to have distinguishable networks [see Eq. (2.60)], e.g. for $\alpha = 0.001$ and $q = 0.75$ we consider network sizes up to $N = 135000$. Again, $p = 10^{-4}$ and $\alpha = 10^{-4}$ are the smallest values considered in our numerical simulations.

In Fig. 2.8 the results are depicted. Unlike below the lines, we observe a mean-field synchronization transition with standard mean-field exponents above them. In particular, the markers correspond to the smallest p or α values for which we observe a synchronization transition with the critical mean-field exponents $\beta = \frac{1}{2}$ and $\nu = 2$ [68]. Those p and α values are denoted by p_{mf} or α_{mf}.

As expected, for sparser networks the mean-field approximation breaks down. Furthermore, it can be seen that random shortcuts favor the mean-field behavior, since we find (roughly) the scaling relations $p_{\mathrm{mf}} \sim 10^{-3q}$ and $\alpha_{\mathrm{mf}} \sim 10^{-7q}$.

We also underline that our approach works, if the first two moments of the degree distribution exist. So, generally, scale-free networks that display a power-law decay $P(k) \sim k^{-\gamma}$ are excluded from our approximation. In [57] it was numerically shown that for $\gamma < 3$ the mean-field approach yields significant deviations. It is further analytically derived in [42] that only for $\gamma > 5$ the order parameter fulfills a finite-size scaling with the standard mean-field exponenents, and for $2 < \gamma < 3$ the exponents depend on the degree exponent γ. Hence, more homogeneous degree distributions that can be characterized by the first two moments, favor the mean-field treatment as it was the case in our small-world example, see Fig. 2.3.

2.6. Networks with degree-dependent frequency dispersion

The interplay between dynamics and network topology has been and remains an important topic, as impressively reviewed in [30, 44, 59–61]. We address here the question, how correlations between connectivity and dynamics on the microscopic level affect the macroscopic behavior of a network. Such a connection between the numbers of links in a network and the functional ability of the node dynamics should be evident in general and possibly caused by various reasons, for instance by limited energy supply, restricted space or chemical resources, etc. Let us just note that indeed various types of neurons differ strongly in the typical number of connections and firing rates [72].

We approach this question by revisiting the synchronization transition of Kuramoto phase oscillators in complex networks. The phenomenon of synchronization suits a benchmark by virtue of its importance as a paradigmatic emergence of collective behavior, as outlined for instance in [2, 73].

There are various possibilities how the individual oscillation frequencies and degrees can be correlated, including correlations between the mean values or the widths of the corresponding distributions. In [74,75] the frequencies and degrees are assumed to be positively correlated, such that mean values and widths of the frequency and degree distributions directly depend on each other. It is found that with increasing positive correlation, the oscillators are easier to synchronize. Gómez-Gardeñes *et al.* recently studied in detail the special case where the natural frequencies of the oscillators are identified with their degrees, i.e. $\omega_i = k_i \; \forall \; i$ [76]. Their main finding is that implementing this type of degree-frequency correlation in scale-free networks leads to an abrupt synchronization transition, resembling a first-order phase transition. This is remarkable, because the synchronization transition was always found to be of second order, if one only considers how different network topologies affect the dynamics. Furthermore, by interpolating between Erdös-Rényi random networks and scale-free networks, Gómez-Gardeñes *et al.* found that the first-order nature of the synchronization transition appears only in scale-free networks, which are characterized by an unlimited dispersion of the degrees. Hence, it was shown that a positive correlation between the dynamics of the oscillators and the large heterogeneity in the number of connections has a drastic effect on the onset of synchronization.

We remark that with respect to many real-world systems, one cannot observe a direct correlation between individual dynamics and connectivity. In neuronal networks for instance, a higher number of connections is not necessarily linked to a higher neuronal firing rate with respect to one cell type. This is partially due to the balance of inhibition and excitation [77], where de- and acceleration compensate each other. Here we take this into account in the form of a rather general correlation between the degree and the frequency distributions; we relate the *diversity* of the frequencies to the degrees. Two different settings are generated, namely either positive or negative correlations between the degree of a node and how broad its oscillatory frequency varies from the mean one. We explore whether this kind of correlation with fixed average natural frequency is sufficient to yield a notable impact on the synchronization transition. In particular, we focus on the Erdös-Rényi random network model, which often serves as an important benchmark [72].

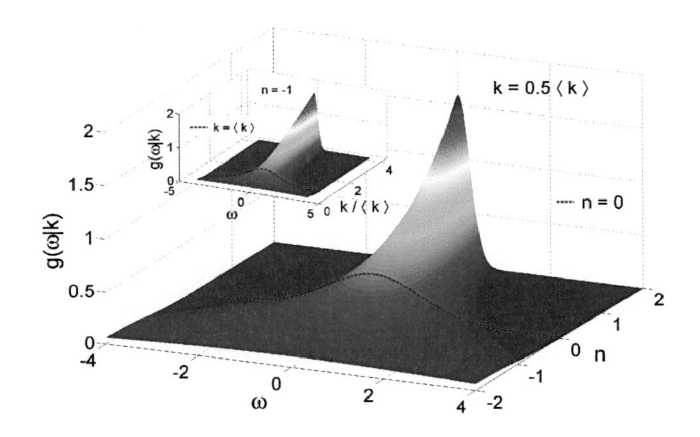

Figure 2.9.: Conditional Gaussian frequency distribution $g(\omega|k)$ with $\sigma_0 = 1$ shown as a function of the correlation power n and the relative degree $k/\langle k \rangle$ (inset), as it follows from Eq. (2.65). From [78].

2.7. Implementation of an example with first numerical experiments

Let us first repeat the variant of the Kuramoto model [15] that we consider here:

$$\dot{\phi}_i(t) = \omega_i + \frac{K}{N^q} \sum_{j=1}^{N} A_{ij} \sin\left(\phi_j(t) - \phi_i(t)\right) + \xi_i(t). \tag{2.62}$$

This equation differs from (2.1) only by the denseness parameter q. As in Sec. 2.5, we use this parameter to guarantee that, even if the overall number of links scales differently with the system size N, the coupling term remains intensive, i.e. does not scale with N. For a sparse network a proper choice is $q = 0$, while for a dense network one takes $q = 1$. Such a normalization is appropriate as long as all the degrees share the same scaling with the system size. Otherwise one may choose the maximum degree occurring in the network to guarantee an intensive coupling term [30]. Again we consider undirected and unweighted networks, in which case the adjacency matrix is symmetric with elements $A_{ij} = 1$, if the units i and j are coupled, otherwise $A_{ij} = 0$.

Assuming a given degree distribution $P(k)$, the joint frequency-degree distribution (cf Sec. 2.3) separates as

$$P(\omega, k) \equiv g(\omega|k)P(k), \tag{2.63}$$

with the conditional frequency distribution $g(\omega|k)$. The latter denotes the probability that an oscillator at a node with given degree k has the natural frequency ω. The central idea in this work is to consider correlations that are due to or affect only the variability in the degree or the frequency distribution, respectively, without affecting the mean values. To this end, we assume that each oscillator draws its natural frequency from the same distribution function, but with an individual standard deviation depending on the degree k_i:

$$\sqrt{\langle \omega_i^2 \rangle - \langle \omega_i \rangle^2} = \sigma_n(k_i). \tag{2.64}$$

Here, $\sigma_n(k_i)$ is an abbreviation for the power-law function

$$\sigma_n(k_i) = \sigma_0 \left(\frac{k_i}{\langle k \rangle} \right)^n, n \in \mathbb{R}. \tag{2.65}$$

We call n the correlation power; σ_0 stands for the original standard deviation without correlations. Eq. (2.65) gives rise to two different settings, namely either positive, $n > 0$, or negative correlations, $n < 0$. Note that for $k > \langle k \rangle$, i.e. nodes with degrees larger than the average degree, the natural frequencies are more broadly distributed around the mean frequency for $n > 0$. In other words, hub nodes then receive a penalty for their many connections by having to sample their natural frequency from a broader distribution. For $k < \langle k \rangle$ or $n < 0$ we have the opposite case. See Fig. 2.9 for a visualization of what we have just explained in words.

Here we consider Erdös-Rényi-like random networks that are constructed by assigning an edge probability

$$p_e = p \cdot N^{q-1}, \ 0 \leq p, q \leq 1 \tag{2.66}$$

for any two of the N nodes in the network [for the scaling parameter q, see Eq. (2.62)]. The further additional requirement is that besides the edge probability, each node is a priori connected to another randomly chosen one. In this way we guarantee that there are no isolated nodes, which are not interesting here, because they are not able to take part in the synchronization process and only reduce the effective system size. Hence, the average degree reads

$$\langle k \rangle = 2 + pN^q \left(1 - \frac{3}{N}\right), \tag{2.67}$$

which is approximately pN^q for $q > 0$ and $N \to \infty$. The first term in Eq. (2.67) stems from the random connections that are a priori chosen, whereas the second term is a result of the edge probabilities, Eq. (2.66). Higher moments and the degree distribution $P(k)$ are not known exactly, but for large systems and $q > 0$, the second term in Eq. (2.67) dominates and the degrees become binomially distributed [60]. To begin with, we construct Erdös-Rényi random networks with edge probability p [$q = 1$ in Eq. (2.66)]. The degrees follow a binomial distribution with the following mean and standard deviation:

$$\langle k \rangle_{\text{binom}} = Np, \quad \text{std}_{\text{binom}}(k) = \sqrt{Np(1-p)}. \tag{2.68}$$

We generate networks of different degree heterogeneity by rewiring the edges and accepting the configurations only if the new degree standard deviation exceeds the original one, $\text{std}(k) > \text{std}_{\text{binom}}(k)$, up to a desired value. The mean degree remains unchanged thereby.

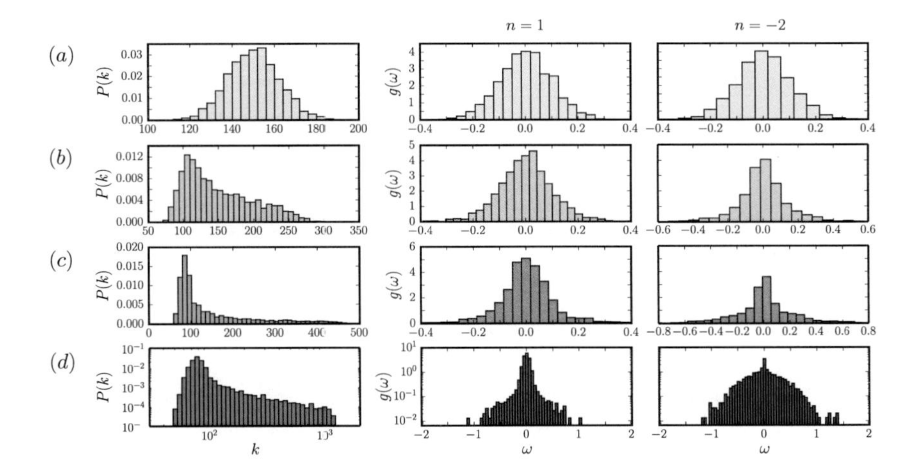

Figure 2.10.: How degree distributions change upon increasing the degree standard deviation $\mathrm{std}(k)$ and how this changes the effective frequency distributions [Eq. (2.69)], for two different correlation powers as indicated in the titles. Gaussian $g(\omega|k)$ with $\sigma_0 = 0.1$ [see Fig. 2.9 and Eq. (2.65)] and Erdös-Rényi random networks of $N = 3000$ nodes with $q = 1$ and $p = 0.05$ [cf. Eq. (2.66)], which leads to $\langle k\rangle_{\mathrm{binom}} = 150$ and $\mathrm{std}_{\mathrm{binom}}(k) \approx 12$, see Eq. (2.68). (a) $\mathrm{std}(k) = \mathrm{std}_{\mathrm{binom}}(k)$, (b) $\mathrm{std}(k) = 4\ \mathrm{std}_{\mathrm{binom}}(k)$, (c) $\mathrm{std}(k) = 8\ \mathrm{std}_{\mathrm{binom}}(k)$, (d) $\mathrm{std}(k) = 16\ \mathrm{std}_{\mathrm{binom}}(k)$.

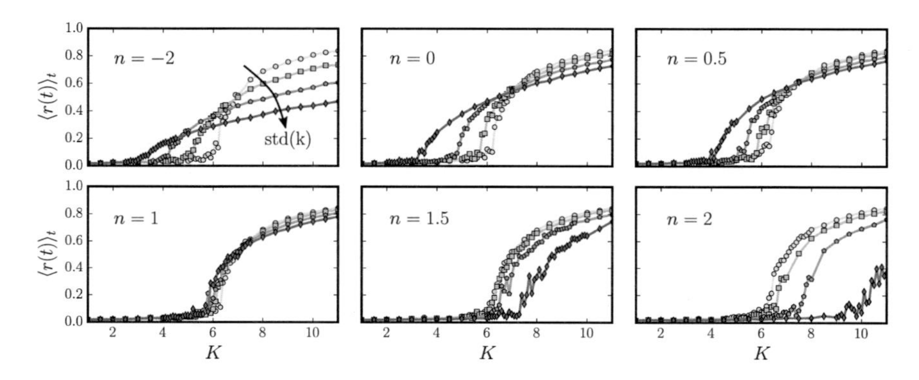

Figure 2.11.: Simulation results showing time-averaged Kuramoto order parameter as a function of coupling strength for different degree standard deviations and different correlation parameters, as indicated in the figure. Lines connect the markers to guide the eye. Single network realizations are used with noise intensity $D = 0.1$. Remaining parameters are the same as in Fig. 2.10).

In Fig. 2.10 we plot degree and effective frequency distributions [cf. Eq. (2.63)]

$$g(\omega) = \sum_{k'} g(\omega|k')P(k') \tag{2.69}$$

for four different network configurations by using Eq. (2.65). As expected, for higher degree standard deviation std(k) the degree distribution becomes more heavy-tailed. Depending on whether the correlations between degree and frequency dispersion are positive $n > 0$ or negative $n < 0$, the effective distribution of all natural frequencies gets a different shape, but the difference is less evident than for the conditional frequency distribution, cf. Fig. 2.9. We observe a different curvature off the mean value $\omega = 0$, most distinguished in Fig. 2.10(d).

As a first test of how the degree-dependent frequency dispersion (2.65) affects the synchronization, we run the dynamics on the the aforementioned networks. We integrate (2.62) until $t = 800$ with time step $\Delta t = 0.05$ by using the Heun scheme [66, 67] and calculate the Kuramoto order parameter time-averaged in $t \in [400, 800]$. Intriguingly, there seems to be a positive correlation $n = n_c > 1$ at which the onset of synchronization is located at the same critical coupling strength, irrespective of the degree heterogeneity in the network, see Fig. 2.11. We observe that below that correlation power, $n < n_c$, it is easier to partially synchronize networks with larger degree heterogeneity, while the opposite is true for $n > n_c$.

2.8. Detailed analysis and clarification of noise effects

We aim to get further insights by combining extensive numerical experiments with an approximate theory. In the numerical simulations, the stochastic differential equations (2.62) are now integrated up to $t = 600$ with time step $\Delta t = 0.05$. Moreover, we discard the data up to $t = 200$, by which transient effects are avoided. We consider a Gaussian frequency distribution with zero mean and standard deviation $\sigma_n(k_i)$, where $\sigma_0 = 0.2$ [cf. Eq. (2.65)], and we use more general Erdös-Rényi random networks with $q = 0.4$ [cf. Eq. (2.66)]. The statistical equilibria are further calculated as averages over at least 100 different network realizations which differ in the configuration of the connections as well as in the oscillators' natural frequencies and initial phase values. In order to estimate the critical coupling strength from numerical simulations, we perform again a finite-size scaling analysis (see Sec. 2.4), where we take networks of sizes $N = 300, 500, 800, 1200$ with $q = 0.4$.

We collect the data in Fig. 2.13 which displays the measured critical coupling strengths (red circles). As expected, both a larger noise intensity D and a decrease of the number of connections, here parameterized by the inverse edge probability $1/p$, impedes the synchronizability. This causes the higher coupling strength K_c needed for the onset of synchronization [compare panels (a)-(d) in Fig. 2.13].

Besides that, we observe that K_c increases with the correlation power n, which is not trivial. One could have expected that both settings of correlation, i.e. $n < 0$ and $n > 0$ in (2.65), lead to a decrease of K_c, because the latter marks the transition from the completely asynchronous to a partially synchronous state, and not to the completely synchronous state. For any correlation power n, oscillators with a narrower frequency distribution appear which are easier to synchronize. Hence, one could argue that a lower coupling strength would be needed for the onset of synchronization.

Instead, the uncorrelated case $n = 0$ has a critical coupling strength intermediate to the two settings with $n \neq 0$. Positive correlations require higher critical coupling strengths K_c, negatively correlated networks can be easier synchronized, i.e. K_c decays.

An intuitive explanation for this observation is that the phenomenon of synchronization arises by virtue of interactions. Therefore, nodes with a larger degree $k > \langle k \rangle$ (hubs) are more crucial than nodes with a smaller degree $k < \langle k \rangle$ (compare with Ref. [79]). If the frequencies of these hubs are much broader spread around the average frequency, it is more difficult for the whole network to exhibit a synchronized oscillation. In other words, a population of oscillators is easier to synchronize, if the important nodes possess frequencies closer to the average frequency. In particular, for the case $n > 0$ hubs are favored to have a great variability of frequencies, whereas sparsely linked nodes draw their frequencies from a sharper distribution [see Fig. 2.9]. Then, the necessary coupling for the onset of synchronization has to be larger than in the uncorrelated case. Differently for $n < 0$, the less linked nodes own an increased variabilty compared to the uncorrelated case, but the hubs are now easier to synchronize since their frequencies are narrower distributed.

We further observe in Fig. 2.13 that the curves for different n approach each other with increasing noise intensity D. This is due to the fact that a strong noise outweighs the diversity of the oscillators given by the frequency distribution; correlations are diminished

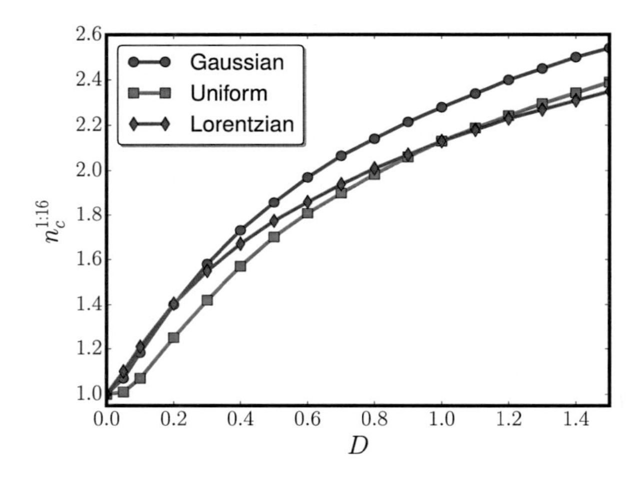

Figure 2.12.: Two numerically generated Erdös-Rényi random networks are considered that were used in Figs. 2.10 and 2.11 [$q = 1$ in Eq. (2.66)]. One has degree standard deviation $\mathrm{std}(k) = \mathrm{std}_{\mathrm{binom}}(k) \approx 12$, the other one $\mathrm{std}(k) = 16\ \mathrm{std}_{\mathrm{binom}}(k)$. The degree distributions for the two networks are inserted into Eq. (2.74) and the correlation power n is varied until the two networks show the same critical coupling strength. The resulting special correlation value is denoted by $n_c^{1:16}$ to indicate the different degree standard deviations in the two networks. Different frequency distributions are used with $\sigma_0 = 0.1$ [cf. Eqs. (2.64),(2.65)].

for large noise intensities and ultimately become negligible.

It turns out to be beneficial to plot the critical coupling strength K_c as a function of the inverse edge probability $1/p$ as done in Fig. 2.13. In this way we observe two distinct regions: for small correlation power n, the critical coupling strength increases sublinearly as a function of $1/p$, while for large n, it increases superlinearly. In panel (d) a linear dependence is located between $n = 1$ and $n = 2$ for $D = 0$. For larger noise intensities or smaller standard deviations σ_0 [compare inset in panel (c)], the separation between the two regions appears for larger correlation powers n. The shaded areas (orange) in panels (c) and (d) depict up to which n we find the superlinear region by numerical simulations.

A linear dependence indicates that the onset of synchronization K_c solely depends on the mean degree $\langle k \rangle$, which is determined by p. Hence, for a certain n_c, the onset of synchronization seems to become independent of higher moments of the degree distribution. In this sense, the heterogeneity in the network is masked by the correlations.

In what follows, we try to reproduce analytically our observations above with the help of an approximate mean-field theory which is based on the the coarse-graining method described in Sec. 2.1. As a result, the linear stability of the completely asynchronous state is characterized by a single real-valued eigenvalue λ given by a self-consistent equation [see Sec. 2.3]:

$$1 = \frac{K}{2N^q \langle k \rangle} \int_{-\infty}^{+\infty} d\omega' \sum_{k'} \frac{(\lambda + D)k'^2}{(\lambda + D)^2 + \omega'^2} P(\omega', k') \ . \tag{2.70}$$

The sum over k' covers all possible degrees, which could be further approximated by an integral. The joint probability density $P(\omega, k)$ takes into account the possibility of correlations between the frequencies and degrees. In the derivation of Eq. (2.70) we assume that, with regard to the ω-dependency, $P(\omega, k)$ has a single maximum at frequency $\omega = 0$ (this choice is always possible due to the rotational symmetry) and is symmetric with respect to it.

The critical condition $\lambda = \lambda_c = 0$ yields the critical coupling strength

$$K_c = 2N^q \langle k \rangle \left[\int_{-\infty}^{+\infty} d\omega' \sum_{k'} \frac{Dk'^2}{D^2 + \omega'^2} P(\omega', k') \right]^{-1} \ . \tag{2.71}$$

This equation is not valid in the noise-free case, where one has to take the limit $\lambda \to 0^+$ in Eq. (2.70) with $D = 0$ resulting in

$$K_c = 2N^q \langle k \rangle \left[\pi \sum_{k'} k'^2 P(0, k') \right]^{-1} \ . \tag{2.72}$$

To see this, note that $\lim_{\lambda \to 0^+} \int_{-\infty}^{+\infty} \mathrm{d}\omega' \lambda / (\lambda^2 + \omega'^2) = \pi \int_{-\infty}^{+\infty} \mathrm{d}\omega' \delta(\omega')$ [54].

First, in accordance with the numerics above, we use Eq. (2.63) with a Gaussian frequency distribution,

$$g_{\mathrm{gauss}}(\omega|k) = \frac{1}{\sqrt{2\pi}\sigma_n(k)} \mathrm{e}^{-\frac{1}{2}\frac{\omega^2}{\sigma_n(k)^2}}, \tag{2.73}$$

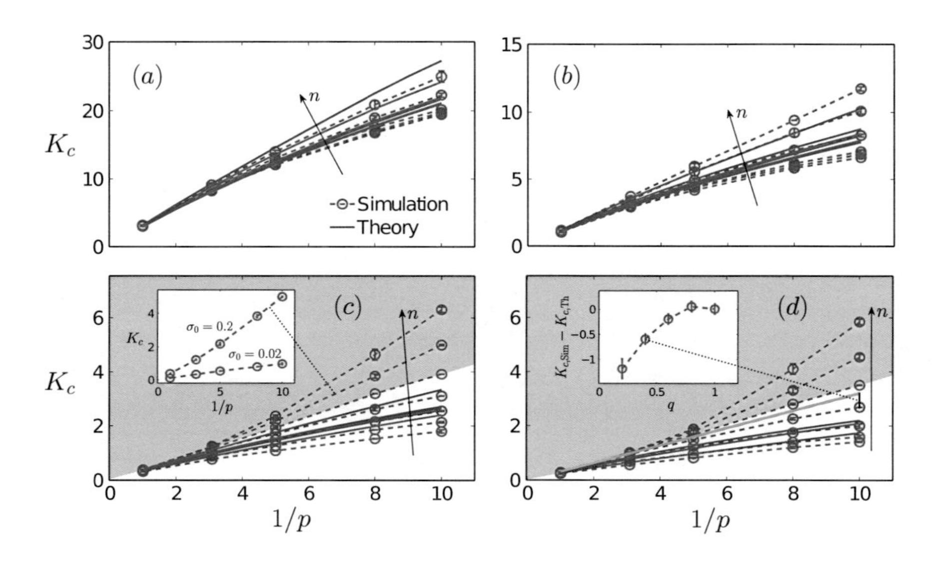

Figure 2.13.: The critical coupling strength depicted as a function of the inverse edge probability, which is proportional to the inverse average degree, see Eq. (2.67). Measurements are done via finite-size scaling analysis with systems of size $N = 300, 500, 800, 1200$ and $q = 0.4$. Furthermore, a Gaussian frequency distribution is used with zero mean and standard deviation $\sigma_n(k_i)$, where $\sigma_0 = 0.2$ [cf. Eq. (2.65)]. Markers connected by dashed lines (red) show simulation results, while (blue) solid lines depict corresponding theoretical results, obtained by numerical integration. The thick solid (green) line in panel (d) is exactly obtained by Eq. (2.76) with $n = 1, 2$. The arrows show the direction of increasing n, from $n = -2$ to $n = 4$. Panels (a)-(d) show results with decreasing noise intensity: (a) $D = 1.5$, (b) $D = 0.5$, (c) $D = 0.05$, and (d) $D = 0$. Inset in (c) presents K_c comparing $\sigma_0 = 0.2$ with $\sigma_0 = 0.02$ in case of $n = 3$. Inset in (d) depicts the discrepancy $K_{c,\mathrm{Sim}} - K_{c,\mathrm{Th}}$ between simulation and theory as a function of the denseness parameter q in case of $n = 1$ and $p = 0.1$. In the unshaded regions of (a)-(d) simulation results are accompanied by theoretical results, whereas in the shaded areas (orange) in (c) and (d) a superlinear growth of K_c cannot be described by our mean-field theory. Modified from [78].

where $\sigma_n(k)$ is expressed by (2.64). Taking the integral, we derive the critical coupling strength (2.71):

$$K_{c,\text{gauss}} = 2\sqrt{\frac{2}{\pi}}\sigma_0 N^q \langle k \rangle^{1-n} \left\langle k^{2-n} \operatorname{erfc}\left(\frac{D}{\sqrt{2}\sigma_n(k)}\right) \exp\left(\frac{D^2}{2\sigma_n(k)^2}\right)\right\rangle^{-1}, \tag{2.74}$$

which is an intensive parameter and scales with the variation of frequencies for small σ_0 [see inset in panel (c) of Fig. 2.13].

By calculating dK_c/dn, we want to validate that the critical coupling strength indeed grows with the correlation power, as stated in the previous section. To this end, we use Eq. (2.74) and arrive at a necessary condition for $dK_{c,\text{gauss}}/dn > 0$, namely

$$\frac{2}{\sqrt{\pi}} y > \left(2y^2 - 1\right) \exp\left(y^2\right) \operatorname{erfc}(y), \tag{2.75}$$

with $y = D\left(\langle k \rangle / k\right)^n / \left(\sqrt{2}\sigma_0\right)$. Since we have $y > 0$, the inequality (2.75) is true; in fact, the right-hand side divided by y, approaches $2/\sqrt{\pi}$ from below for y going to infinity.

In the noise-free case $D = 0$ we find

$$K_{c,\text{gauss}}(D = 0) = 2\sqrt{\frac{2}{\pi}}\sigma_0 N^q \frac{\langle k \rangle^{1-n}}{\langle k^{2-n} \rangle}. \tag{2.76}$$

For $n = 1$ and $n = 2$ we get the same critical coupling strength growing inversely with the average degree:

$$K_{c,\text{gauss}}(D = 0, n = 1, 2) = 2\sqrt{\frac{2}{\pi}}\frac{\sigma_0}{p} \tag{2.77}$$

in the thermodynamic limit [cf. Eq. (2.67)]. In Fig. 2.13 the solid blue lines describe the critical coupling strength as given by numerical integration of Eqs. (2.74) or (2.76) with a binomial degree distribution and system size $N = 1000$. In panel (d) for $n = 1, 2$, instead of the numerical integration of Eq. (2.76), the exact expression (2.77) is shown (thick green line). The agreement between theory and simulation is satisfactory in (a)-(d) and confirms the previous observations. It seems that we cannot find superlinear dependencies within the mean-field approximation. Simulation results in the shaded (orange) areas in panels (c) and (d) are therefore not covered by the theory; the validity of the approximation restricts to correlation strengths with sub- and linear growth of K_c. For large noise intensities D the theory overestimates the critical coupling strength K_c, irrespective of the correlation power n, but this may turn into the opposite case when decreasing D depending on n. In summary, there seems to be some particular noise value where the agreement between theory and simulation is particularly good. As presented in the inset of panel (d), deviations between the results from the numerical simulations $K_{c,\text{Sim}}$ and from the weighted mean-field theory $K_{c,\text{Th}}$ can be further reduced by increasing q. Hence, more densely connected networks are better reflected by the theory, in agreement with our former observations in Sec. 2.5. It appears that the masking of degree heterogeneity for $n = 2$ in Eq. (2.77) is a spurious solution that results from the mean-field approximation not valid anymore in that regime.

The network model under consideration constitutes already a generalized model, since it allows to interpolate between sparse and dense random networks. Let us discuss two further generalizations, namely other frequency distributions and different normalization variants of the coupling term, cf. Eq. (2.62). In particular, we consider now a Lorentzian and a uniform frequency distribution:

$$\begin{aligned} g_{\text{lorentz}}(\omega|k) &= \frac{\sigma_n(k)}{\pi}\frac{1}{\sigma_n(k)^2+\omega^2}, \\ g_{\text{uni}}(\omega|k) &= \frac{1}{2\sqrt{3}\sigma_n(k)}, \quad |\omega| \le \sqrt{3}\sigma_n(k), \end{aligned} \tag{2.78}$$

Note that in case of the Lorentzian, $\sigma_n(k)$ does not have the meaning of a standard deviation, instead it is the scale parameter for the width of the distribution. We further introduce a generalized normalization $\mathcal{N}(k)$ instead of N^q, which can be a function of the degree k. Then we find for the critical coupling strength:

$$\begin{aligned} K_{c,\text{lorentz}} &= 2\langle k\rangle \left\langle \frac{k^2}{\mathcal{N}(k)}\frac{1}{D+\sigma_n(k)}\right\rangle^{-1}, \\ K_{c,\text{uni}} &= 2\sqrt{3}\sigma_0\langle k\rangle^{1-n}\left\langle \frac{k^{2-n}}{\mathcal{N}(k)}\arctan\left(\frac{\sqrt{3}\sigma_n(k)}{D}\right)\right\rangle^{-1}. \end{aligned} \tag{2.79}$$

Let us now specify the normalization $\mathcal{N}(k)$ by considering two cases: $\mathcal{N}(k) = \langle k\rangle$ and $\mathcal{N}(k) = k$. In the first case, one assumes again that the system-size scaling of the number of connections is the same for all nodes. In order to distinguish the two normalizations, we denote the critical coupling strength by K_a or K_w, respectively. In the noise-free case $D = 0$ we obtain, in contrast to (2.76), the following relations with the same constant of proportionality C:

$$K_a = C\sigma_0\frac{\langle k\rangle^{2-n}}{\langle k^{2-n}\rangle}, \quad K_w = C\sigma_0\frac{\langle k\rangle^{1-n}}{\langle k^{1-n}\rangle}, \tag{2.80}$$

irrespective of whether we consider a Gaussian, a Lorentzian or a uniform frequency distribution. Only the constant of proportionality C is different, namely $C_{\text{gauss}} = 2\sqrt{2/\pi} \approx 1.60$, $C_{\text{lorentz}} = 2$ or $C_{\text{uni}} = 4\sqrt{3}/\pi \approx 2.21$.

Again we find a disappearance of network effects for specific correlation powers. The phenomenon is even more pronounced here, since K_a becomes a constant for $n = 1, 2$. Moreover we see that $K_w(n) = K_a(n+1)$. Correspondingly, K_w becomes a constant for $n = 0, 1$. It has been pointed out in the literature, e.g. in [30], that the additional weight introduced by the normalization $\mathcal{N}(k) = k$ can mask the heterogeneity of the network. Our results constitute a generalization of this statement. Preliminary numerical simulations can reproduce our theoretical result $K_w(n) = K_a(n+1)$, while the point where the onset of synchronization loses its dependence on the network connectivity is found to appear at smaller values of n than predicted by the theory.

2.9. Summary and outlook

We have investigated noisy Kuramoto phase oscillators on undirected ring networks. Analytical tractability was achieved within a heterogenous mean-field approximation valid for random uncorrelated networks. The complex network structure was essentially approximated by a weighted fully connected network, where the weights are obtained through a combinatorial consideration of the connectivity purely based on the degrees. This procedure allowed us to write down the nonlinear Fokker-Planck equation for the evolution of the one-oscillator phase distribution and to analytically perform the linear stability analysis of the incoherent state. As a result, we obtained the critical coupling strength that marks the onset of synchronization in the network.

First, we continued with the case where degrees and natural frequencies do not depend on each other. Then the critical coupling was found to be a product of two functionals. One of them was a functional of the degree distribution and therefore depended solely on the network topology, while the other one was a functional of the frequency distribution and a function of the noise intensity. This result did coincide with previous results [54,80], except of a rescaling by the ratio of the first two moments of the degree distribution. In particular, achieving partial synchrony is impeded by noise and frequency mismatch, but facilitated by adding connections. We observed that the latter is particularly pronounced in the small-world network regime.

In order to compare the theoretical results with numerical simulations, we have introduced a dense small-world network model. As typical of small-world networks [65], we started with a regular network consisting only of next neighbor connections, and then we randomly addded shortcuts. We found that increasing the network size makes regular local edges and random shortcuts more similar in their effects. A good agreement between analytical and numerical results was obtained. Specifically, finite-size scaling analysis with standard mean-field exponents always gave a well-defined intersection point such that the Kuramoto model on dense small-world networks showed a synchronization transition of standard mean-field type.

We performed additional numerical experiments, where we interpolated between sparsely and densely connected networks and in which we considered separately purely random shortcuts and purely regular local couplings. The results obtained thereby suggested that random connections favor a mean-field approach, because the regular networks had to be more densely connected in order to yield synchronization transitions with standard mean-field exponents.

Next we considered an example for correlations between the degrees and the dispersion of natural frequencies. By numerically and analytically estimating the onset of synchronization, we were able to show that correlations can favor as well as impede the ability to generate coherent network oscillations. In both scenarios, the behavior of hubs played a dominant role. Stronger coupling was necessary, when the hubs had more broadly distributed frequencies. The onset of synchonization was shifted up, even though the less linked nodes at the same time drew their frequencies from a narrower distribution, such that taken separately, they would synchronize at lower couplings. The opposite happened when hubs had narrowed distributions.

Most interestingly, we found that for certain positive correlations the network structure could be masked such that networks with different degree distributions yield the same critical coupling strength. Remarkably, a similar phenomenon was independently observed in [81].

Analytical results agreed satisfactorily with numerical simulations for sufficiently large noise intensity D or not too strong correlation powers n and dense enough scaling of the links $q > 0$, in particular, as long as the critical coupling grew sub- or linearly with $1/p$. Beyond a certain denseness parameter for given correlation power, the weighted mean-field approximation broke down.

The noise effects can be summarized as follows. In Sec. 2.3 we have shown how noise impedes synchronization, in particular we have seen that the incoherent state is neutrally stable if the noise vanishes, $D = 0$, but becomes linearly stable for $D > 0$. This generalizes, at least in an approximate way, the findings in [54] towards random networks. In Sec. 2.8 we clarified that also in case of degree-frequency correlations the noise-free case, $D = 0$, is singular, because only then the shape of the frequency distribution does not enter.

Highly heterogeneous networks in which the mean-field description breaks down, have also been discussed.

3. Approximate solution to the stochastic Kuramoto model

In the last chapter we were concerned about the transition from incoherence to partial synchronization among noisy oscillators. While this is a very distinct point, one would like to have more analytical insights into the collective dynamics that takes place in such populations. This opens up a topic that has been explored for a long time and where still many open questions remain, namely the low-dimensional behavior that evidently hides behind the N-dimensional Kuramoto model (often the infinite system size limit $N \to \infty$ is considered) [82–94]. Probably the most remarkable recent result is due to Ott and Antonsen [90, 91]. They discovered that a power-law ansatz can be used to express all Fourier coefficients of the one-oscillator probability density by the very same complex function. As a consequence, one can find drastic but still exact dimensionality reductions for a broad class of such systems. The Ott-Antonsen ansatz relies crucially on quenched disorder in the natural frequencies, i.e. a time-independent frequency distribution $g(\omega)$ of non-zero variance[1]. This excludes uncorrelated noise acting on the instantaneous frequencies[2]. While the proper counterpart of the Ott-Antonsen ansatz remains to be found in the stochastic domain, it is possible to obtain reduced systems with low-dimensional dynamics by virtue of an approximation.

3.1. Exposition of the Gaussian approximation

The idea is to separate the whole oscillator network into subpopulations of the same individal quantities, cf. Eq. (1.1), and then to assume that the corresponding one-oscillator probability distributions in the thermodynamic limit $N \to \infty$ (see Sec. 2.2) are Gaussians with time-dependent means and variances. Such an ansatz is motivated by numerical observations [98, 99], but we will make clear in this chapter that the Gaussian ansatz can also be motivated from a theoretical point of view. As a result, one can derive low-dimensional reduced systems, where

$$\text{number of ODE's} = 2 \times \text{number of subpopulations.} \tag{3.1}$$

It is beneficial to start with the most simple scenario of all-to-all coupled identical Kuramoto oscillators, where the diversity purely comes from noise acting on the frequencies. Since for this system exact results exist, we can systematically assess the accuracy of the

[1] For identical frequencies the Ott-Antonsen manifold is only neutrally stable [87].

[2] Common noise is admissible within the Ott-Antonsen theory [95–97].

Gaussian approximation. So the model we focus on here reads:

$$\dot{\phi}_i(t) = \omega_0 + \frac{K}{N}\sum_{j=1}^{N} \sin(\phi_j - \phi_i) + \xi_i(t). \tag{3.2}$$

All oscillators have the same constant natural frequency ω_0 concerning the underlying limit-cycle. By virtue of the rotational symmetry in the model, we can subtract the natural frequency from all instantaneous frequencies without changing the dynamics. This co-rotating frame is equivalent to setting $\omega_0 = 0$ in Eq. (3.2), without loss of generality. The stochastic forces $\xi_i(t)$ perturb the evolution of the phases. Such time-dependent disorder is often modeled by Gaussian white noise [22, 100], which we also consider here. Therefore we have $\langle \xi_i(t) \rangle = 0$, $\langle \xi_i(t)\xi_j(t') \rangle = 2D\delta_{ij}\delta(t-t')$, where the angular brackets denote averages over different realizations of the noise and the nonnegative parameter D scales its intensity. As mentioned before, the noise terms $\xi_i(t)$ can be regarded as an aggregation of various stochastic processes, such as the variability in the release of neurotransmitters or the quasi-random synaptic inputs from other neurons. One can set D or K to unity (by rescaling time), but for illustrative purposes we keep the two parameters in Eq. (3.2). Since all oscillators are identical in (3.2), the system can be described by a single one-oscillator probability density $\rho(\phi, t)$ in the thermodynamic limit $N \to \infty$, see Sec. 2.2. From (3.1) we expect that the dimensionality of the whole system can be reduced to two ODE's by assuming that $\rho(\phi, t)$ is a Gaussian distribution. We recall the normalization condition $\int_0^{2\pi} \rho(\phi, t) d\phi = 1 \; \forall \; t$; $\rho(\phi, t) \; d\phi$ gives the fraction of oscillators having a phase between ϕ and $\phi + d\phi$ at time t. The nonlinear Fokker-Planck equation for the evolution of the one-oscillator probability density $\rho(\phi, t)$ reads

$$\frac{\partial \rho(\phi, t)}{\partial t} = D\frac{\partial^2 \rho(\phi, t)}{\partial \phi^2} - \frac{\partial}{\partial \phi}\left\{ K r(t) \sin\left[\Theta(t) - \phi\right] \rho(\phi, t)\right\} , \tag{3.3}$$

$$r(t)\mathrm{e}^{i\Theta(t)} = \int_0^{2\pi} d\phi' \; \mathrm{e}^{i\phi'} \; \rho(\phi', t) \; . \tag{3.4}$$

Consider the Fourier series

$$\rho(\phi, t) = \frac{1}{2\pi} \sum_{n=-\infty}^{+\infty} \hat{\rho}_n(t)\mathrm{e}^{-in\phi} \tag{3.5}$$

with $\hat{\rho}_0 = 1$ and $\hat{\rho}_{-n} = \hat{\rho}_n^*$. Through the inverse transform the remaining Fourier coefficients are given by

$$\hat{\rho}_n(t) = \int_0^{2\pi} d\phi' \; \rho(\phi', t) \; \mathrm{e}^{in\phi'} \equiv c_n(t) + is_n(t). \tag{3.6}$$

Here we have introduced the variables c_n and s_n to denote the real and imaginary parts of the Fourier coefficients, respectively. Apparently, $n = 1$ leads to Eq. (3.4); in particular, the classical Kuramoto order parameter equals $r(t) = |\hat{\rho}_1(t)|$. Inserting the Fourier series into (3.4) yields an infinite chain of coupled complex-valued equations for the Fourier coefficients [101]

$$\frac{\dot{\hat{\rho}}_n}{n} = \frac{K}{2}\left(\hat{\rho}_{n-1}\hat{\rho}_1 - \hat{\rho}_{n+1}\hat{\rho}_{-1}\right) - Dn\hat{\rho}_n, \tag{3.7}$$

with $n = 1, 2, \ldots, \infty$. Importantly, the coefficients $\hat{\rho}_n(t)$ decay rapidly with increasing n, see Fig. 3.2. So one can obtain an approximate description of the underlying dynamics with arbitrary accuracy by truncating Eqs. (3.7) at a large enough n. Typically, $n = 6$ leads already to results of satisfying accuracy. Here we have however a different focus, because we aim for an approximate way to close the infinite chain (3.7). This result can then be compared with a large-n truncation of (3.7).

Specifically, we approximate the phase distribution $\rho(\phi, t)$ by a time-dependent wrapped Gaussian. In general, a wrapped distribution can be written in terms of the characteristic function φ_l of the unwrapped distribution as [102]

$$\rho(\phi, t) = \frac{1}{2\pi} \sum_{l=-\infty}^{\infty} \varphi_l(t)\, \mathrm{e}^{-il\phi}. \tag{3.8}$$

Comparing with Eq. (3.5) we see that the Fourier coefficients coincide with the characteristic function of the unwrapped distribution, which in case of the Gaussian reads $\varphi_l(t) = \exp[-\sigma^2(t) l^2/2 + ilm(t)]$, with time-dependent mean $m(t)$ and variance $\sigma^2(t)$ of the unwrapped Gaussian. This can be checked by inserting (3.8) into Eq. (3.6):

$$\begin{aligned} \hat{\rho}_n^g(t) &= \frac{1}{2\pi} \int_0^{2\pi} d\phi' \sum_{l=-\infty}^{\infty} \mathrm{e}^{-l^2\sigma^2(t)/2 + i[(n-l)\phi' + lm(t)]} \\ &= \frac{1}{2\pi} \sum_{l=-\infty}^{\infty} \mathrm{e}^{-l^2\sigma^2(t)/2 + ilm(t)} \int_0^{2\pi} d\phi' \mathrm{e}^{i(n-l)\phi'} \\ &= \mathrm{e}^{-n^2\sigma^2(t)/2 + inm(t)}. \end{aligned} \tag{3.9}$$

The superscript g labels expressions obtained within the Gaussian theory. In the last step we notice that the periodic function integrated over its period gives a non-zero contribution only if $l = n$. Hence, real and imaginary parts of $\hat{\rho}_n^g(t)$, Eq. (3.6), turn into

$$c_n^g(t) = \mathrm{e}^{-n^2\sigma^2(t)/2} \cos\left[nm(t)\right], \quad s_n^g(t) = \mathrm{e}^{-n^2\sigma^2(t)/2} \sin\left[nm(t)\right]. \tag{3.10}$$

From Eq. (3.10) it follows that the Fourier amplitudes are not affected by the mean phase:

$$|\hat{\rho}_n^g(t)| = \mathrm{e}^{-n^2\sigma^2(t)/2}. \tag{3.11}$$

Noteworthy, we observe a similarity between the Gaussian ansatz for the Kuramoto model with noise and the Ott-Antonsen ansatz for the case with quenched disorder instead of noise. The latter namely consists of identifying the nth Fourier coefficient with the nth power of a specific complex function [90, 91]. It has to be emphasized that in contrast to the Gaussian ansatz, the Ott-Antonsen ansatz is exact (for limitations see [93] and for a recent extension see [94]). The benefit of the Gaussian ansatz lies essentially in the fact that all higher Fourier coefficients can *explicitly* be deduced from the first one, such that a closure in (3.7) is achieved. For instance, the real and imaginary parts of the second Fourier coefficient are obtained as $c_2^g = (c_1^g)^4 - (s_1^g)^4$ and $s_2^g = 2s_1^g c_1^g\left[(s_1^g)^2 + (c_1^g)^2\right]$, respectively [99]. Thus, the dimensionality of the system is reduced to two coupled ODE's:

$$\begin{aligned} \dot{c}_1^g &= -Dc_1^g + \frac{Kc_1^g}{2}\left\{1 - \left[(c_1^g)^2 + (s_1^g)^2\right]^2\right\}, \\ \dot{s}_1^g &= -Ds_1^g + \frac{Ks_1^g}{2}\left\{1 - \left[(c_1^g)^2 + (s_1^g)^2\right]^2\right\}. \end{aligned} \tag{3.12}$$

Transforming the remaining variables $\{c_1^g, s_1^g\} \to \{m, \sigma^2\}$ yields ODE's for the cumulants

$$\dot{\sigma^2} = 2D + K\left(\mathrm{e}^{-2\sigma^2} - 1\right) \tag{3.13}$$

and $\dot{m} = 0$ (compare with [99]). The problem is now even simpler, as we have to integrate only a single differential equation. This is made possible here, because the mean of the phase distribution is constant in time in the co-rotating frame, otherwise $\dot{m} = \omega_0$, see Eq. (3.2). A useful alternative transformation is the one via polar coordinates, $c_1^g = r^g \cos\Theta^g$ and $s_1^g = r^g \sin\Theta^g$. Then one obtains a differential equation for the mean-field amplitude $r(t)$, which is the classical Kuramoto order parameter:

$$\dot{r}^g = \frac{r^g}{2}\left\{K\left[1 - (r^g)^4\right] - 2D\right\}. \tag{3.14}$$

For the derivative of the mean phase the Gaussian approximation yields $\dot{\Theta}^g = 0$. Therefore, the approximate level of synchronization in the oscillator population shall be obtained by solving a single evolution equation.

3.2. Time-dependent solutions and long-time limits

The differential equation (3.13) can be integrated directly after separation of variables. The variance of the phases then reads

$$\sigma^2(t) = \frac{1}{2}\ln\left[\frac{K}{K-2D} + \mathrm{e}^{-2t(K-2D)}\left(\mathrm{e}^{2\sigma^2(0)} - \frac{K}{K-2D}\right)\right] \tag{3.15}$$

for $K \neq 2D$. Inserting this expression into Eq. (3.8) and making use of the fact that the characteristic function is given by the Fourier coefficients (3.9), we can write down a full solution for the evolution of the phases in terms of a wrapped Gaussian (we set the arbitrary mean $m(t)$ to zero):

$$\rho^g(\phi, t) = \frac{1}{2\pi}\left\{1 + 2\sum_{n=0}^{\infty}\left[\frac{K}{K-2D} + \mathrm{e}^{-2t(K-2D)}\left(\mathrm{e}^{2\sigma^2(0)} - \frac{K}{K-2D}\right)\right]^{-n^2/4}\cos(n\phi)\right\} \tag{3.16}$$

For the long-time limit $t \to \infty$ we label the variables with the subscript 0. We see that for $t \to \infty$ the variance grows to infinity for $K < 2D$. Apparently, this leads to a uniform phase distribution $\rho_0^g(\phi) = 1/2\pi$, which is characteristic for the completely asynchronous state. In contrast, for $K > 2D$ the variance asymptotes to

$$\sigma_0^2 = \frac{1}{2}\ln\left(\frac{K}{K-2D}\right), \tag{3.17}$$

and for the phase distribution we obtain

$$\rho_0^g(\phi) = \frac{1}{2\pi}\left\{1 + 2\sum_{n=0}^{\infty}(1 - 2D/K)^{n^2/4}\cos(n\phi)\right\}. \tag{3.18}$$

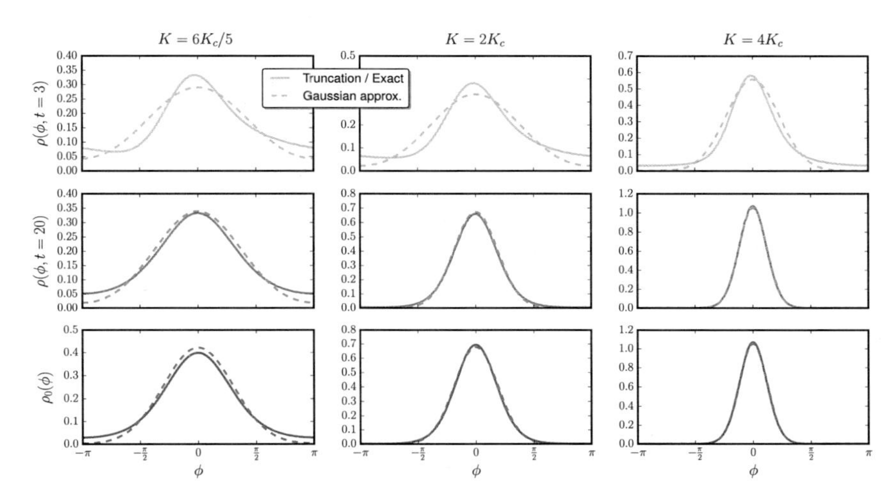

Figure 3.1.: Snapshots of phase distributions. For the theoretical lines in the transient regime the variance from Eq. (3.15) is used. The initial variance is $\sigma^2(0) = 1.96$. For all snapshots, phases are shifted by the same amount so that mean values are located at $\phi = \pi$. The parameters are the same as in Fig. 3.3(a1)'s uppermost curve.

Hence, the variance σ^2 equals zero, either if the coupling strength goes to infinity, $K \to \infty$, or if the noise intensity D vanishes. This situation would correspond to a perfectly synchronized state. In contrast, the completely asynchronous state is characterized by a diverging variance $\sigma^2 \to \infty$, while a finite non-zero variance, $\sigma^2 > 0$, signals partial synchronization. In summary, at $K_c = 2D$ the oscillator population transitions from incoherent to partially synchronized behavior. Noteworthy, this critical value is *exact*, as it is known for a long time [80]. In the sequel, the Fourier amplitudes are of particular interest. They were used already in Eq. (3.16). Now, by inserting (3.15) into (3.11), we can write:

$$|\hat{\rho}_n^g(t)| = \left| \frac{K - K_c}{\left(\frac{K - K_c}{|\hat{\rho}_n^g(0)|^{4/n^2}} - K \right) \mathrm{e}^{-2t(K - K_c)} + K} \right|^{n^2/4} . \tag{3.19}$$

It is remarkable that we can report this approximate solution even though the effects of noise on the collective dynamics of phase oscillators were comprehensively studied before [48, 54, 63, 80, 92, 103–109].

Equation (3.19) indicates that $|\hat{\rho}_n^g(t)|$ goes to zero exponentially fast with increasing time for subcritical coupling, while it goes to

$$|\hat{\rho}_n^g(t \to \infty)| = [1 - (K_c/K)]^{n^2/4} \tag{3.20}$$

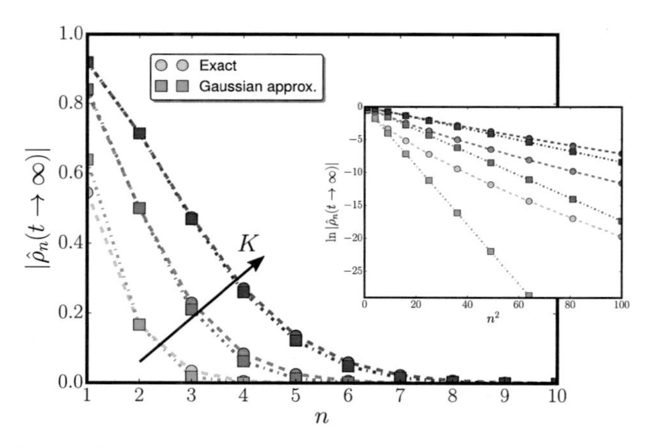

Figure 3.2.: Decay of stationary Fourier amplitudes with order n. For clarity, points are connected by lines. Gaussian theory is compared with exact results (3.7). Coupling strengths are $K = \{1.2K_c, 2K_c, 3.5K_c\}$. The inset shows the same data with different axes.

for $K > K_c$. Note that the solution for the classical Kuramoto order parameter (3.14) follows from Eq. (3.19) by taking $n = 1$:

$$r^g(t) = \left| \frac{K - K_c}{\left(\frac{K-K_c}{[r^g(0)]^4} - K\right) \mathrm{e}^{-2t(K-K_c)} + K} \right|^{1/4} . \tag{3.21}$$

Correspondingly, in the long-time limit $t \to \infty$ and for supercritical coupling strengths, $K > K_c$, the stationary order parameter reads

$$r_0^g = [1 - (K_c/K)]^{1/4} . \tag{3.22}$$

Apparently, Eq. (3.22) violates the well-known square-root scaling law [80]. In the next section we will show that the Gaussian approximation is otherwise very accurate.

3.3. Gaussian theory vs. numerical experiments and exact results

Let us start mentioning exact results for the steady state, which can serve as a benchmark for the Gaussian approximation. It is known that in the long-time limit the phases are exactly distributed according to a von Mises distribution (again we center the distribution around zero without loss of generality),

$$\rho_0(\phi) = \frac{\mathrm{e}^{r_0 K/D \cos\phi}}{2\pi I_0(r_0 K/D)}, \tag{3.23}$$

where the stationary order parameter r_0 is given via a transcendental equation [109]:

$$r_0 = \frac{I_1(r_0 K/D)}{I_0(r_0 K/D)}. \tag{3.24}$$

I_0 and I_1 denote modified Bessel functions of the first kind of order 0 and 1, respectively. Similarly, the higher Fourier amplitudes satisfy

$$|\hat{\rho}_n(t \to \infty)| = \frac{I_n(r_0 K/D)}{I_0(r_0 K/D)}. \tag{3.25}$$

Thus, all higher Fourier amplitudes can be calculated provided the first one, $r_0 = |\hat{\rho}_1(t \to \infty)|$, is found. In this way a closure is in fact an inherent property of the stochastic Kuramoto model. However, we stress that it is not explicit here. Explicitness is realized in the Gaussian approximation, where the closure then even extends into the time-dependent domain. In Fig. 3.1 we compare several phase distributions. For the stationary regime the exact distribution is obtained by finding the zeros in Eq. (3.24) and inserting them into Eq. (3.23), while for the wrapped Gaussian we use (3.18) with a large-n truncation in the sum. On the other hand, in the transient regime we use Eq. (3.5) with a large-n truncation of the Fourier coefficients (3.7) to obtain practically exact results. Similarly we truncate the sum in Eq. (3.16) at some large n to obtain the time-dependent phase distribution in Gaussian approximation. We observe that the phase distribution is well approximated by a wrapped Gaussian, if the coupling strengths are sufficiently large, $K \gtrsim 2K_c$. This can be expected, because the von Mises distribution (3.23) approaches Gaussianity for large arguments, i.e. for $r_0 K/D \gg 1$. We further observe that the exact distributions have heavier tails than the wrapped Gaussians, and this is particularly pronounced in the transient regime and close to the critical coupling strength $K \gtrsim K_c$. The transient regime additionally manifests asymmetry in the phase distribution, which inherently cannot be described by a Gaussian distribution, since by definition the latter has no skewness. The latter point can be improved by imposing the Gaussian distribution as the initial phase distribution.

In Fig. 3.2 we go on to compare for the steady state the analytical result of the Fourier amplitudes (3.20) with the numerical truncation of (3.7). We see that the decay of the Fourier coefficients with the order n is captured by the Gaussian theory. Plotting $\ln|\hat{\rho}_n(t \to \infty)|$ as a function of n^2 should give straight lines. The inset of Fig. 3.2 shows that the Gaussian approximation tends to underestimate the higher Fourier amplitudes. Furthermore, there is a distinct deviation in the vicinity of the critical coupling strength for the first Fourier amplitude, visible for the lowest curve in the main part of Fig. 3.2, which is the main subject in the remainder.

The first Fourier amplitude coincides with the classical Kuramoto order parameter (3.21). In Fig. 3.3 we show its time dependence. While panels (a1) and (a2) show how the analytical result compares with numerical experiments, panel (b) compares the Gaussian approximation with truncation of (3.7) at some n. We observe that the theory overestimates the order parameter, in particular in the non-stationary regime and slightly above the critical coupling K_c. This is consistent with the observations we made in Figs.

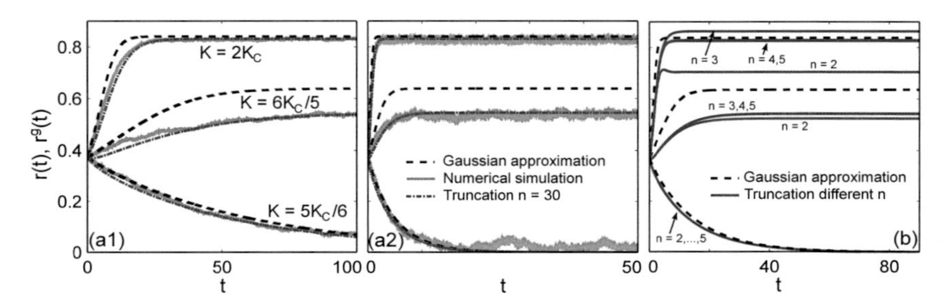

Figure 3.3.: Time evolution of Kuramoto order parameter with initial condition $r(0) = r^g(0) = 0.373$. (a1) compares theory and numerical experiments for different coupling strengths K ($D = 0.1$). The black dashed line shows the Gaussian theory $r^g(t)$, the blue dash-dotted line the numerical integration of truncated Eqs. (3.7) at $n = 30$, and orange solid line the simulation of full dynamics (3.2) with initially Gaussian distributed phases. In (a2) a different noise intensity is used, $D = 1.2$. Simulation of full dynamics is performed with system size $N = 15000$ and time steps 0.05 for $D = 0.1$ and 0.005 for $D = 1.2$, respectively. (b) compares the accuracy between theory (dashed lines) and truncation of (3.7) at some n (solid lines). Modified from [110].

3.1 and 3.2. Below K_c, the theoretical lines are close to the numerically obtained ones. For sufficiently strong coupling strengths the theory clearly outperforms numerical truncation. This can be seen in Fig. 3.3(b) for $K = 2K_c$. Note that $n = 3$ corresponds already to six coupled differential equations, while the Gaussian ansatz leads to a two-dimensional system [effectively one-dimensional, see discussion of Eq. (3.13)].

Finally, we proceed with discussing the long-time behavior of the order parameter as a function of the coupling strength, i.e. we compare Eq. (3.22) with Eq. (3.24). In figure 3.4 panel (a) we compare our scaling result with numerical experiments. In agreement with our observations concerning the time-dependent order parameter (Fig. 3.3), there is a distinct deviation for coupling strengths in the interval $K_c < K \lesssim 2K_c$. This is due to the fact that Eq. (3.22) does not capture the well-known square-root scaling law $r_0 \sim (K - K_c)^{1/2}$ near criticality [80]; for completeness, we show the crossover to the square-root scaling in the inset of Fig. 3.4(a). Noteworthy, the analytical scaling result is highly accurate for sufficiently large coupling strengths, i.e. $K \gtrsim 2K_c$.

Recently, besides other interesting results, Bertini *et al.* proved certain bounds for the asymptotic order parameter, namely $(1 - K_c/K)^{1/2} < r_0 < (1 - K_c/2K)^{1/2}$ [109]. Now, with the general form of Bernoulli's inequality, we find $(1 - K_c/K)^{1/4} < (1 - K_c/2K)^{1/2}$. Indeed, $(1 - K_c/K)^{1/4}$ seems to establish an improved upper bound. This is shown in Fig. 3.4 panel (b), where we depict the bounds proven by Bertini *et al.* in comparison with our scaling result and the "exact" solution, obtained from Eq. (3.24) via numerical determination of the roots. In the inset one can appreciate the accuracy of our result.

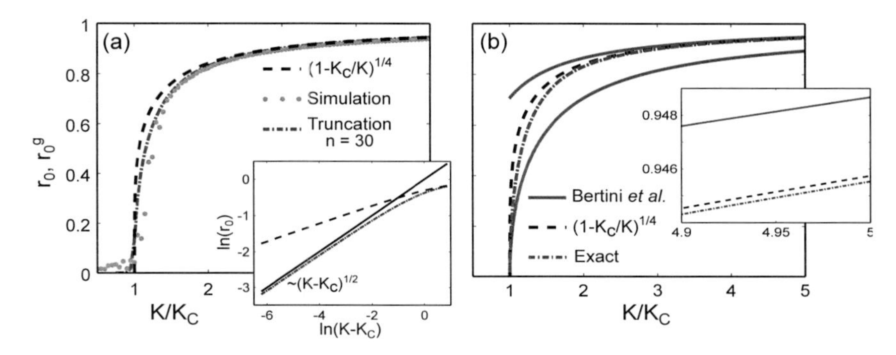

Figure 3.4.: Scaling of the asymptotic order parameter as a function of coupling strength divided by critical value, K/K_c. In (a) theory (3.20) for $n = 1$, r_0^g, is compared with numerical experiments [numerical integration of (3.2) with $N = 5000$, $D = 0.1$, integration step 0.05 and time-averaged between $t = 100$ and $t = 200$; truncated system (3.7) is integrated until $t = 1500$ and the final value for the order parameter is taken]. The inset shows square-root scaling slightly above critical coupling, which is not covered by the theory; the smallest value there $K - K_c = 0.002$. In (b) theory is compared with the exact solution, Eq. (3.24), and with results from [109]. The inset shows a zoom-in. Modified from [110].

3.4. Temporal fluctuations vs. quenched disorder

Kuramoto showed for a general symmetric and unimodal frequency distribution $g(\omega)$ that r_0 satisfies [15]

$$1 = K \int_{-\pi/2}^{\pi/2} \cos(\phi)^2 g\left[r_0 K \sin(\phi)\right] d\phi \, . \tag{3.26}$$

From expanding $g\left[r_0 K \sin(\phi)\right]$ in powers of $r_0 K$, he then found a square-root scaling law for r_0 as $K \to K_c$. Notably, Kuramoto also found that in case of a Lorentzian $g(\omega)$, the order parameter equals exactly $r_0 = \sqrt{1 - K_c/K}$ for all $K > K_c$. Note that for a Gaussian $g(\omega)$ with standard deviation σ, (3.26) can be brought into a form that is similar to (3.24), namely a transcendental equation with the aforementioned Bessel functions:

$$\exp\left(\frac{r_0^2 K^2}{4\sigma^2}\right)\sqrt{\frac{8\sigma^2}{\pi K^2}} = I_0\left(\frac{r_0^2 K^2}{4\sigma^2}\right) + I_1\left(\frac{r_0^2 K^2}{4\sigma^2}\right) \tag{3.27}$$

The comparison between Gaussian and Lorentzian quenched disorder and Gaussian white noise is given in Fig. 3.5. With respect to K/K_c, the Lorentzian quenched disorder hinders synchronization most strongly, then in the middle comes Gaussian white noise, and Gaussian quenched disorder hinders synchronization in the weakest form.

3.5. Including complex networks

Here we present the extension of the Gaussian approximaton towards additional time-independent disorder, by which the oscillators can be discriminated. Let us denote such individual quantity with σ_i [see Eq. (1.1) and Sec. 2.2]. Then the idea is to gather all oscillators with the same individual quantities into one subpopulation, and to describe the oscillators' phases in each subpopulation by a time-dependent Gaussian distribution. The Gaussians then interact with each other through their first two cumulants. Equivalently one can describe each subpopulation by local mean-field variables $r_\sigma(t)$ and $\Theta_\sigma(t)$ and all of those are then coupled to each other. Now we come to complex coupling structures as an example of additional time-independent (quenched) disorder. The problem is analytically intractable in general. The remedy lies in finding a suitable approximative description. To this end, we employ the coarse-graining explained in Sec. 2.1. Let A be the adjacency matrix: $A_{ij} = 1$ if node j couples to node i and $A_{ij} = 0$ if this is not the case. Let further k_i denote the degree, the number of connections of node i. Then for an undirected network where the adjacency matrix is symmetric, we approximate the latter by $\tilde{A}_{ij} = k_i k_j / \sum_l k_l$. As a result, all nodes with the same degree k build a subpopulation with their own mean-field variables $r_k(t), \Theta_k(t)$:

$$r_k(t)\mathrm{e}^{i\Theta_k(t)} = \int_0^{2\pi} d\phi' \, \mathrm{e}^{i\phi'} \, \rho\left(\phi', t|k\right). \tag{3.28}$$

The difference to the all-to-all uniformly coupled case [Eq. (3.4)] appears in the one-oscillator probability density $\rho\left(\phi, t|k\right)$, which now depends on the individual degree k. The subpopulations are coupled through averages over the degree distribution $P(k)$, such that the global mean-field variables are given by

$$r(t)\mathrm{e}^{i\Theta(t)} = \langle k' r_{k'}(t) \mathrm{e}^{i\Theta_{k'}(t)}\rangle / \langle k' \rangle. \tag{3.29}$$

Here we use the notation $\langle \ldots \rangle \equiv \sum_{k'} \ldots P(k')$. In Gaussian approximation we find

$$\begin{aligned}
\dot{r}_k^g &= \frac{1-\left(r_k^g\right)^4}{2N\langle k' \rangle} \, K \, k \, \langle k' r_{k'}^g \cos\left(\Theta_{k'}^g - \Theta_k^g\right)\rangle - r_k^g D, \\
\dot{\Theta}_k^g &= \frac{\left(r_k^g\right)^{-1} + \left(r_k^g\right)^3}{2N\langle k' \rangle} \, K \, k \, \langle k' r_{k'}^g \sin\left(\Theta_{k'}^g - \Theta_k^g\right)\rangle.
\end{aligned} \tag{3.30}$$

We may set $\Theta_k^g(t) = 0$ in the co-rotating frame. Then the order parameter becomes $r^g = \langle k' r_{k'}^g \rangle / \langle k' \rangle$. Instead of Eq. (3.14) one obtains

$$\dot{r}^g = \frac{K}{2N\langle k' \rangle} \left\langle k'^2 \left[1 - \left(r_{k'}^g\right)^4\right]\right\rangle r^g - D r^g. \tag{3.31}$$

Near the synchronization transition $r_k \to 0 \ \forall k$, so the contributions of $\left(r_k^g\right)^4$ can be neglected. Therefore, at $K_c = 2DN\langle k' \rangle / \langle k'^2 \rangle$ the order parameter switches from exponentially decreasing to increasing. Again, the Gaussian approximation reproduces the known critical coupling strength, compare with Eq. (2.41).

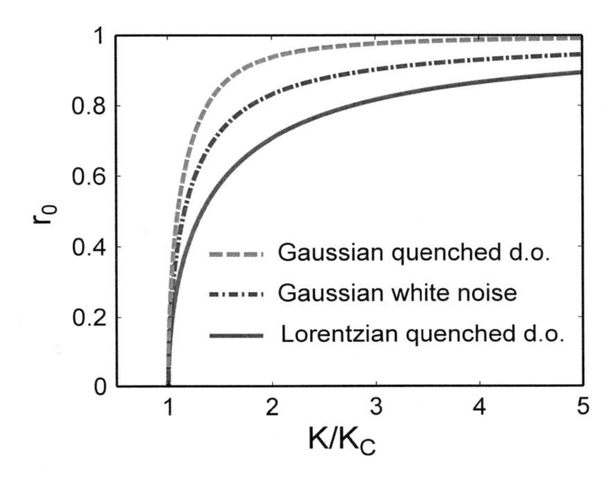

Figure 3.5.: Long-time asymptotic levels of synchronization as a function of coupling strength divided by critical value, K/K_c, for Gaussian and Lorentzian quenched disorder and Gaussian white noise. Modified from [110].

3.6. Summary and outlook

We have revisited the Kuramoto model where the only source of disorder comes from temporal fluctuations acting on the evolution of the oscillator phases. By assuming Gaussianity in the phase distribution at all times, we found an approximate reduced description in the continuum limit, when the system is in fact infinitely dimensional. In the Gaussian approximation the system consists of two uncoupled ordinary first-order differential equations, and therefore the problem effectively turned out to be even one-dimensional. As a consequence, we could easily find a full solution. Specifically, we derived an expression for the time-dependent order parameter, which approximately revealed the level of synchronization for any point in time and for any coupling strength. The critical coupling strength for the onset of synchronization is exactly reproduced by the theory. We also found that the Gaussian approximation is accurate for a coupling weaker or twice as strong as the critical one. In the vicinity of the critical value, the Gaussian theory did not reproduce the square-root scaling. Remarkably, however, the obtained scaling law appeared to be an improved explicit upper bound for the stationary order parameter. It was also interesting to see that the Gaussian approximation provides a simple way to calculate analytically the synchronization transition point in complex networks. By showing where the Gaussian ansatz is valid, and where it is incorrect (and to what extent), this chapter shed new light on the underlying low-dimensional dynamics. One promising line of research for the future would be to search for a class of models that allow the Gaussian theory to be exact. Then one could potentially obtain a unified and more complete picture of the low-dimensional manifolds that can govern the mutual synchronization of interacting elements.

4. Excitable elements controlled by noise and network structure

Here we go one step further by allowing excitability as another mode of individual dynamics. Excitable elements are ubiquitous in physical, geophysical, chemical and biological systems. Examples range from chemical reactions, neuronal systems, cardiovascular tissues and climate dynamics to laser devices, see Refs. [22, 100] and references therein. An excitable system possesses a stable equilibrium, but, once excited by sufficiently strong perturbations, displays large non-monotonic excursions before returning to the state of rest.

Networks of excitable elements are omnipresent in nature, and the human brain, as a gigantic network of neurons, appears to be paradigmatic in this respect (cf. [9]). The basic question then is, whether the coupled excitable elements can together overcome an excitation threshold and leave the resting state, which can be seen as the cooperative surmounting of a bottleneck [111]. If the excitation threshold is surpassed with a lack of cooperation, the resulting collective state will be incoherent.

Phase oscillators in their canonical form are not directly applicable to studies of coupled excitable elements. In particular, a Kuramoto oscillator possesses translational invariance with respect to phase shifts: it rotates uniformly, and all phase values are dynamically equivalent, whereas for an excitable element a certain event (e.g. spike) is singled out. An "active rotator" – a modification of the Kuramoto model which takes into account these properties of excitable elements – was introduced by Shinomoto and Kuramoto in [112]. Since then, the analysis was refined and certain additional aspects were investigated in [95, 108, 113–115]. Most of the research considered the case of uniform global coupling in the ensemble; only in a few works, active rotators were studied on a network with complex structure, see e.g. [116]. Here we study dynamics of networks of randomly connected stochastic active rotators and put emphasis on how temporal fluctuations and the network structure influence the dynamics of the network as a whole.

4.1. Noise-driven active rotators

In the population of noise-driven active rotators, introduced by Shinomoto and Kuramoto [112], the dynamics of individual phases $\phi_i(t)$ is governed by

$$\dot{\phi}_i(t) = 1 - a \sin[\phi_i(t)] + \frac{K}{N} \sum_{j=1}^{N} A_{ij} \sin[\phi_j(t) - \phi_i(t)] + \xi_i(t), \qquad (4.1)$$

The new parameter a determines the excitation threshold of each separate element. This model essentially consists of a set of Adler's equations [117] with Kuramoto coupling. Noteworthy, the Adler's equation itself is still routinely used to investigate how oscillators are affected by external signals, see e.g. [118–120].

We restrict ourselves first to the case where all elements of the ensemble have the same natural frequency $\omega_0 = 1$ and coupling strength K. The network is assumed to be undirected and unweighted, hence the elements of the symmetric adjacency matrix A adopt only two values: $A_{ij} = 1$, if the units i and j are coupled, and $A_{ij} = 0$ otherwise. Complex topologies of real-world networks can be encoded into the adjacency matrix, and decoded by counting all the degrees, which gives rise to a degree distribution $P(k)$.

The terms $\xi_i(t)$ in (4.1) are statistically independent across the network nodes and are zero mean Gaussian white noise sources, $\langle \xi_i(t) \rangle = 0, \quad \langle \xi_i(t)\xi_j(t') \rangle = 2D\delta_{ij}\delta(t-t')$, where the angular brackets denote averages over different realizations of the noise and D is the noise intensity.

For an isolated element in the absence of noise, $\dot{\phi}_i = 1 - a\sin(\phi_i)$, excitable behavior is apparent. For $|a| > 1$ the stable equilibrium is at $\phi_i^\infty = \arcsin(a^{-1})$, and the system needs a sufficiently strong perturbation in order to make an excursion around the circle. Noise plays this role driving the system to escape from the equilibrium state ϕ_i^∞. An escape event corresponds to the release of a single spike [22, 98, 112]. For $|a| < 1$ the element shows oscillatory behavior with frequency $\sqrt{1-a^2}$. Notably, evolution of the phase ϕ is not uniform in this case: it is at the slowest near $\phi = \pi/2$ and at the fastest near $\phi = 3\pi/2$.

We proceed with the coarse-grained system as explained in Sec. 2.1, which is valid if individual parameters are not correlated with each other. Then all nodes in the network are fully characterized by their degrees, and a mean-field description can be constructed. Specifically, we study the nonlinear FPE (2.18) for the one-oscillator probability density $\rho(\phi, t|k)$, Eq. (2.23),

$$\frac{\partial \rho(\phi,t|k)}{\partial t} = -\frac{\partial}{\partial \phi}\left[v_k(\phi,t)\,\rho(\phi,t|k)\right] + D\frac{\partial^2 \rho(\phi,t|k)}{\partial \phi^2}\,. \tag{4.2}$$

In contrast to the previous situations, the drift $v_k(\phi, t)$, Eq. (2.19), now contains an extra term with the excitation threshold a,

$$v_k(\phi,t) = 1 - a\sin[\phi(t)] + r(t)KkN^{-1}\sin[\Theta(t) - \phi]\,, \tag{4.3}$$

$$r(t)e^{i\Theta(t)} = \frac{\langle r_{k'}(t)\ k'\ e^{i\Theta_{k'}(t)}\rangle_{k'}}{\langle k' \rangle}, \tag{4.4}$$

$$r_k(t)e^{i\Theta_k(t)} = \int_0^{2\pi} d\phi'\ e^{i\phi'}\rho(\phi',t|k). \tag{4.5}$$

The averages $\langle \ldots \rangle_{k'} \equiv \sum_{k'} \ldots P(k')$ also appear in the evolution equations for the Fourier coefficients. Namely, for every degree k that appears in the network, the Fourier coefficients $\rho_n(t|k)$ are governed by the following infinite chain of coupled complex-valued equations [compare with Eq. (3.7)]:

$$\begin{aligned}\frac{\dot{\hat\rho}_n(t|k)}{n} =& \frac{\tilde{K}k}{\langle k'\rangle}\left(\hat\rho_{n-1}(t|k)\sum_{k'}\hat\rho_1(t|k')k'P(k') - \hat\rho_{n+1}(t|k)\sum_{k'}\hat\rho_{-1}(t|k')k'P(k')\right)\\ &+ \frac{a}{2}\left[\hat\rho_{n-1}(t|k) - \hat\rho_{n+1}(t|k)\right] - (Dn - i)\hat\rho_n(t|k),\end{aligned} \tag{4.6}$$

with the abbreviation $\tilde{K} := K/(2N)$. The mean degree $\langle k'\rangle$ of the network is the first moment of the degree distribution. As expected, in the case of all-to-all connectivity the results of [101] are recovered. Equation (4.6) provides a complete description of our system in the mean-field approximation, which is valid as long as $k \gg 1$. Thus, for a sparsely connected network the mean-field description is likely to fail [57, 63, 121]. We also note that the above mean field approach is restricted to networks with finite second moment of the degree distribution [42, 57, 63].

A closure of the infinite set of equations (4.6) can be achieved within the Gaussian approximation (GA) as explained in Sec. 3.1. It is assumed that in every subset of oscillators with the same network degree, the distribution of the phases at every moment in time can be described by a time-dependent wrapped Gaussian.

As in Sec. 3.5 the dimension of the reduced system of equations for the Fourier coefficients is twice the number of different degrees present in the network. Denoting $c_1(t|k)$ by c_1 and $s_1(t|k)$ by s_1, Eqs. (4.6) turn into

$$\begin{aligned}\dot c_1 =& \frac{\tilde{K}k}{\langle k'\rangle}\left[\langle c_1(t|k')k'\rangle(1 - c_1^4 + s_1^4) - 2\langle s_1(t|k')k'\rangle s_1c_1(s_1^2 + c_1^2)\right]\\ &+ \frac{a}{2}(1 - c_1^4 + s_1^4) - Dc_1 - s_1,\\ \dot s_1 =& \frac{\tilde{K}k}{\langle k'\rangle}\left[\langle s_1(t|k')k'\rangle(1 + c_1^4 - s_1^4) - 2\langle c_1(t|k')k'\rangle s_1c_1(s_1^2 + c_1^2)\right]\\ &- as_1c_1(s_1^2 + c_1^2) - Ds_1 + c_1,\end{aligned} \tag{4.7}$$

for every degree k in the network. Equivalently, equations for the mean $m_k(t)$ and the variance $\sigma_k^2(t)$ are

$$\begin{aligned}\dot m_k =& 1 - \mathrm{e}^{-\sigma_k^2/2}\cosh\sigma_k^2\left[a\sin m_k - \frac{Kk}{N\langle k'\rangle}\langle k'\mathrm{e}^{-\sigma_{k'}^2/2}\sin(m_{k'} - m_k)\rangle\right],\\ \dot\sigma_k^2/2 =& D - \mathrm{e}^{-\sigma_k^2/2}\sinh\sigma_k^2\left[a\cos m_k + \frac{Kk}{N\langle k'\rangle}\langle k'\mathrm{e}^{-\sigma_{k'}^2/2}\cos(m_{k'} - m_k)\rangle\right].\end{aligned} \tag{4.8}$$

4.2. First example: regular networks

The integro-differential character of Eqs. (4.8) makes further general analysis cumbersome. Below we discuss explicitly two relatively simple examples of regular and binary random networks. For regular networks, where all nodes have the same degree, Eqs. (4.8) simplify to two coupled equations:

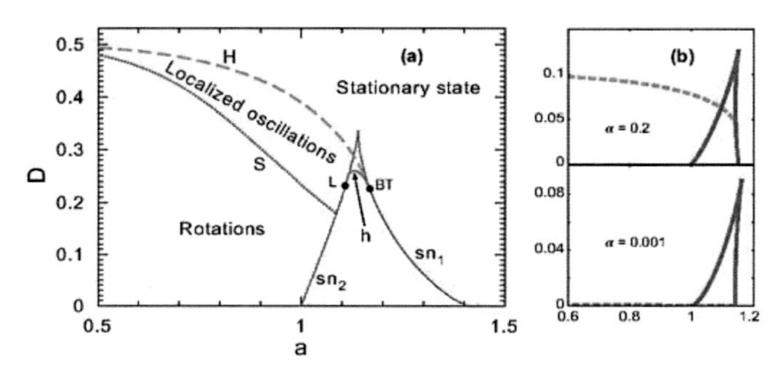

Figure 4.1.: Bifurcation diagram of a regular network of stochastic rotators in the GA. (a): The coupling strength equals $K = 1$ and the connectivity fraction α is unity. $sn_{1,2}$: saddle-node bifurcation lines; H: Hopf bifurcation line; h: homoclinic bifurcation, S: oscillation type changes; BT: Bogdanov-Takens bifurcation point, L: homoclinic loop attached to the saddle-node. (b): Fragments of the bifurcation diagram for the indicated values of the connectivity fraction α. From [122]. A similar version of panel (a) appeared in [99].

$$\dot{m}_k = 1 - a\mathrm{e}^{-\sigma_k^2/2}\cosh\sigma_k^2 \sin m_k, \quad \dot{\sigma}_k^2 = 2D - 2\mathrm{e}^{-\sigma_k^2/2}\sinh\sigma_k^2\left[a\cos m_k + KkN^{-1}\mathrm{e}^{-\sigma_k^2/2}\right]. \tag{4.9}$$

The only difference to the known cumulant equations for a fully connected network [99], i.e. $k = N$, is a rescaling of the coupling strength, namely $K \to Kk/N$, compare panels (a) and (b) in Fig. 4.1. Figure 4.1 panel (a) shows a complete bifurcation diagram for all-to-all coupling on the parameter plane excitability a and noise intensity D. Two saddle-node bifurcations, denoted by sn_1 and sn_2, separate parameter regions in which only a single steady state exists from the central wedge-shaped region in which three steady states are present. While at sn_2 a saddle point and a node are created, sn_1 destroys the saddle and the original steady state. On both bifurcation curves one of the Jacobian eigenvalues vanishes. There is a point where the second eigenvalue vanishes as well. This codimension-2 Bogdanov-Takens (BT) bifurcation is the origin of two further bifurcation lines: the Hopf bifurcation H and the homoclinic bifurcation h. While above H, the steady state is stable, below H the stable limit cycle comes into existence. The curve h marks the existence of an orbit homoclinic to the saddle point. In the parameter region between the curves H, h and sn_2 the stable steady state coexists with the attracting limit cycle. In the region between H and $sn_{1,2}$ two stable steady states exist. At the point L, saddle-node and homoclinicity merge together [99].

Above the Hopf bifurcation line H, individual units are firing incoherently, and therefore do not create a globally synchronized oscillation. So in that stationary state, even though m_k and σ_k are constant in time, the current $\sum_j \langle\dot{\phi}_j(t)\rangle_t/N$ does not vanish ($\langle\ldots\rangle_t$ stands for time average); it rather grows with further increase of D until saturation [108]. Below

H, small-amplitude oscillations are born, where m_k is localized between two phase values ("localized oscillations"). With decreasing noise intensity D their amplitude grows. Below the line S, oscillations turn into full rotations: instead of oscillating back and forth, the ensemble rotates around the whole circle $[0, 2\pi)$. Upon the line S, σ_k becomes infinite; below S it is finite again. Finally, below the saddle-node bifurcation line, the system reaches again a stationary state, but this time the oscillators are mostly at rest without producing any current $\sum_j \langle \dot{\phi}_j(t) \rangle_t / N \sim 0$.

In the following, we do not discriminate between localized oscillations and full-circle rotations. We focus on how oscillating patterns are affected by the network structure. Therefore, we concentrate on finding the Hopf and the saddle-node bifurcations.

Since a reduction in the number of connections effectively decreases the coupling strength, no new qualitative phenomena are expected if all nodes share the same degree. In what follows, we use the abbreviation $\alpha := k/N$ for the connectivity fraction.

We detect numerically the dependence on α of Hopf and saddle-node bifurcations in Eqs. (4.9) for different excitation thresholds a with the help of the software package `MATCONT` [123]. Results are summarized in Fig. 4.2 showing the bifurcation value of noise intensity (called the critical noise intensity) D_c versus α for fixed values of the excitability parameter a. We observe that both the Hopf and the saddle-node bifurcations are shifted to smaller noise intensities when the connectivity is decreased. Dependence of the Hopf threshold on α is practically linear, in accordance with the expansion

$$D_c = K\alpha \left(\frac{1}{2} - \frac{3}{32}a^4 - \frac{3}{256}a^8 + \dots \right). \tag{4.10}$$

As one would expect, the oscillatory regions shrink for higher values of threshold a.

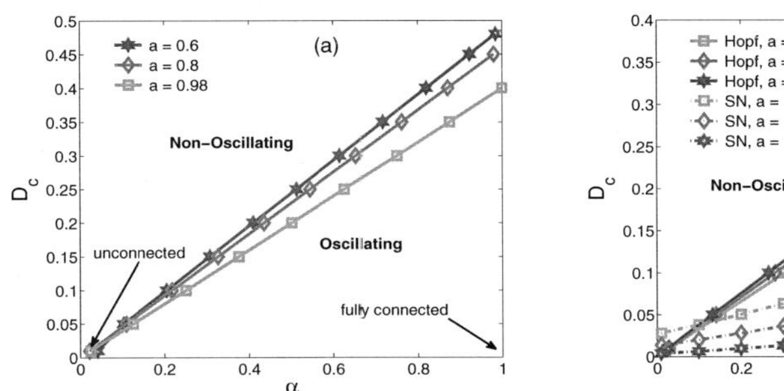

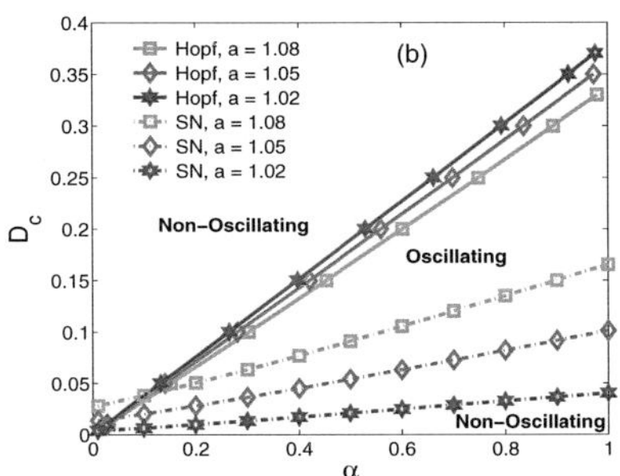

Figure 4.2.: Critical noise intensity D_c as a function of the connectivity fraction α. (a): For low and moderate threshold a, only the Hopf bifurcations are present. (b): Large values of a with Hopf and saddle-node (SN) bifurcations present. Parameter values are indicated in the legends. The coupling strength is $K = 1$ in both panels. From [122].

4.3. Second example: binary random networks

A simple example of a complex heterogeneous network is a binary random network which possesses two distinct connectivity degrees. Denoting these as k_1 and k_2 ($\alpha_1 = k_1/N$, $\alpha_2 = k_2/N$), the degree distribution is given by

$$P(k) = p\delta_{k,k_1} + (1-p)\delta_{k,k_2}. \tag{4.11}$$

We observe that these networks tend to be disassortative in the sense that nodes with different degrees are preferentially connected [124]. Such degree-degree correlations could be factored in by developing a theory along the lines of Chap. B. It turns out (not shown here) that the disassortativity attains large values, if the variance among the connections is large. The maximum variance that can be obtained with (4.11) is VarMax $= 0.25$. We consider only values up to $\mathrm{Var}(\alpha)/\mathrm{VarMax} = 0.5$, where the disassortativity is negligible.

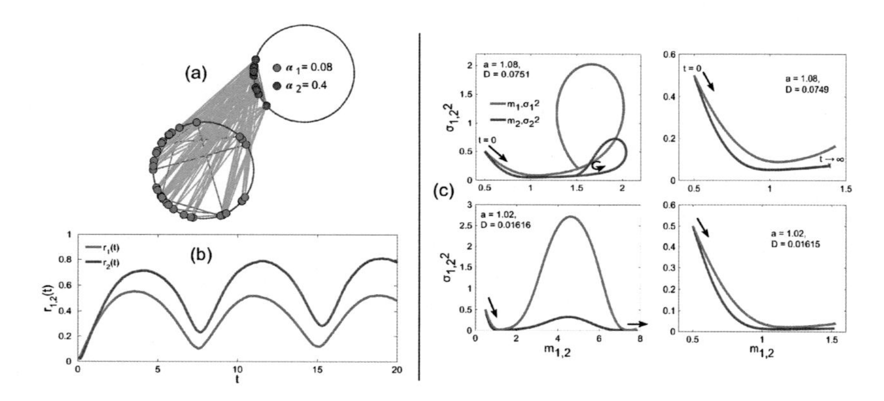

Figure 4.3.: Dynamics of a binary random network of stochastic rotators. (a): A snapshot at $t = 100$ of a network composed of 50 elements. For clarity, the nodes are distributed on two separated rings according to their degrees and phases. Blue nodes have degree $k_1 = 4$, red ones have $k_2 = 20$. Green links are chosen between different degrees, while the orange lines connect nodes with the same degree. Remaining parameters are $a = 0.2$, $D = 0.1$, $K = 2$, $p = 0.7$. (b): Order parameters $r_1(t) = |\rho_1(t|k_1)|$ and $r_2(t) = |\rho_1(t|k_2)|$ obtained from simulation of the *full dynamics*, Eq. (4.1). The parameters are: $N = 10^4$, $k_1 = 1000$, $k_2 = 4000$, $p = 0.7$, $a = 0.6$, $D = 0.1$, $K = 1.5$. (c): Phase portraits of the GA Eqs. (4.12) for $\alpha_1 = 0.37$, $\alpha_2 = 0.97$, $p = 0.95$, $K = 1$ and initial conditions $m_{1,2}(t=0) = \sigma^2_{1,2}(t=0) = 0.5$. Upper left panel depicts localized oscillations, lower left panel depicts full-circle rotations. In the right panels a small decrease of noise has the result that the system approaches fixed points (right endpoints of the trajectories). From [122].

Inserting the degree distribution (4.11) into Eqs. (4.8) gives a system of four coupled first-order differential equations:

$$\begin{aligned}
\dot{m}_1 =& 1 - \mathrm{e}^{-\sigma_1^2/2}\cosh(\sigma_1^2)\left[a\sin(m_1) + \frac{K\alpha_1\alpha_2(1-p)}{p\alpha_1+(1-p)\alpha_2}\sin(m_1-m_2)\mathrm{e}^{-\sigma_2^2/2}\right]\\
\dot{\sigma}_1^2 =& 2D - 2\mathrm{e}^{-\sigma_1^2/2}\sinh(\sigma_1^2)\\
&\times\left[a\cos(m_1) + \frac{K\alpha_1}{p\alpha_1+(1-p)\alpha_2}\left(\alpha_1 p\mathrm{e}^{-\sigma_1^2/2} + \alpha_2(1-p)\mathrm{e}^{-\sigma_2^2/2}\cos(m_1-m_2)\right)\right]\\
\dot{m}_2 =& 1 - \mathrm{e}^{-\sigma_2^2/2}\cosh(\sigma_2^2)\left[a\sin(m_2) + \frac{K\alpha_1\alpha_2 p}{p\alpha_1+(1-p)\alpha_2}\sin(m_2-m_1)\mathrm{e}^{-\sigma_1^2/2}\right]\\
\dot{\sigma}_2^2 =& 2D - 2\mathrm{e}^{-\sigma_2^2/2}\sinh(\sigma_2^2)\\
&\times\left[a\cos(m_2) + \frac{K\alpha_2}{p\alpha_1+(1-p)\alpha_2}\left(\alpha_2(1-p)\mathrm{e}^{-\sigma_2^2/2} + \alpha_1 p\mathrm{e}^{-\sigma_1^2/2}\cos(m_2-m_1)\right)\right].
\end{aligned} \tag{4.12}$$

Bifurcation analysis of this system was performed using `MATCONT` [123].

Fig. 4.3(a) illustrates a binary random network. In (b) we show the order parameters $r_{1,2}(t)$ (3.29) obtained by simulation of the full dynamics (4.1). More connections appear to increase the similarity between the phases in the corresponding population. Fig. 4.3(c) shows phase portraits of the GA Eqs. (4.12) near the saddle-node bifurcation. The left panels depict the oscillatory regime. Indeed, the set of nodes with more connections is more synchronized both in the case of localized oscillations (upper left) and for full rotations (lower left). This is signalled by smaller variances $\sigma_2^2(t)$. In the right panels of (c) a tiny decrease in the noise intensity brings the whole system into a stationary state, where the variables $m_{1,2}$, $\sigma_{1,2}^2$ approach a fixed point.

In this context we never observed that one population of a certain degree is oscillating, while the other is not. This might be the result of entanglement of two populations with different connectivity degrees. Hence, a visualization such as in Fig. 4.3(a) has to be taken with care.

In order to isolate the effects of network heterogeneity, we fix the average degree, $\langle\alpha\rangle = 0.4$. The variance of α is tuned by varying three parameters: α_1, α_2 and p. Since $\mathrm{Var}(\alpha)$ is not uniquely defined and different parameter sets yield the same variance, we take the path where the α-values are the largest, because then the mean-field assumption is best fulfilled. Fig. 4.4 shows critical noise intensity D_c at which a Hopf or saddle-node bifurcations occur in the GA system (4.12) versus the connectivity variance $\mathrm{Var}(\alpha)/\mathrm{VarMax}$. In the oscillatory regime, $a < 1$, shown in Fig. 4.4(a), D_c vs $\mathrm{Var}(\alpha)/\mathrm{VarMax}$ shows trends similar to the case of $a = 0$ [63],

$$D_c(a=0) = \frac{K}{2N}\frac{\langle k'^2\rangle}{\langle k'\rangle} = \frac{K}{2}\left(\langle\alpha\rangle + \frac{\mathrm{Var}(\alpha)}{\langle\alpha\rangle}\right). \tag{4.13}$$

Nonlinear dependence of D_c vs $\mathrm{Var}(\alpha)$ starts to appear only near the threshold $a = 1$. In the excitable regime $a > 1$ [Fig. 4.4(b)] the oscillatory region grows with the increase of $\mathrm{Var}(\alpha)$, if the variance is already large. In particular, the Hopf bifurcations shift to larger

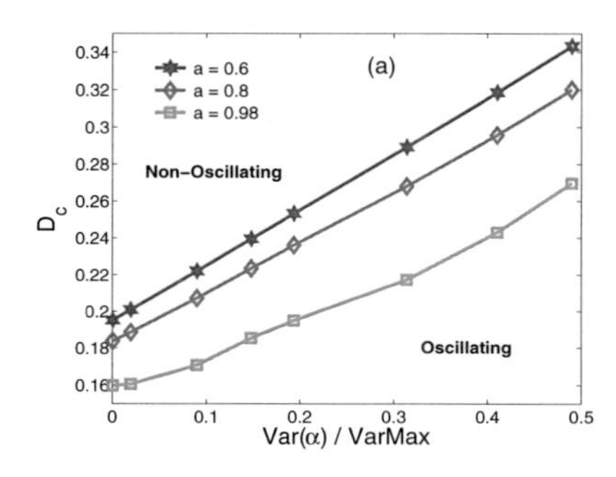

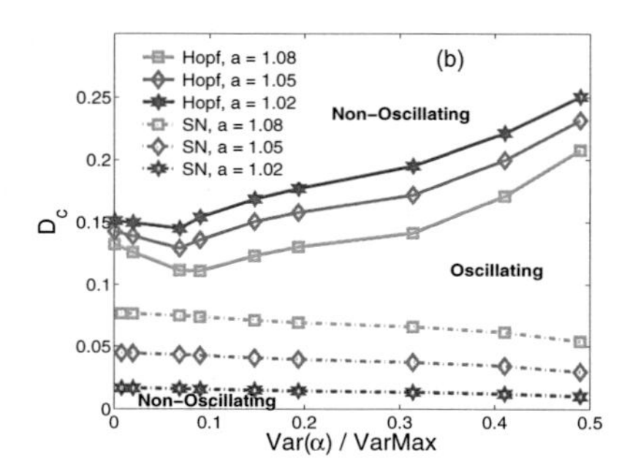

Figure 4.4.: Critical noise intensity D_c versus $\mathrm{Var}(\alpha)/\mathrm{VarMax}$ obtained in the GA for the binary random network of stochastic rotators. As in Fig. 4.2 panels (a) and (b) show the oscillatory and the excitatory regimes, respectively. The values of excitability parameter a are indicated in the legends along with the type of bifurcation: Hopf or saddle-node (SN). Other parameters are $\mathrm{VarMax} = 0.25$, $\langle\alpha\rangle = 0.4$, $K = 1$. From [122].

noise values, while the saddle-node bifurcations shift to smaller ones. Furthermore, for the Hopf bifurcation the dependence D_c vs $\mathrm{Var}(\alpha)$ becomes non-monotone with minima at moderate values of the structural heterogeneity. These minima become more pronounced for larger values of a. This effect constitutes a qualitative difference to the oscillatory regime, $a < 1$, and is induced by the network structure.

In particular, the minima observed for the Hopf bifurcation lines result from adding a small fraction of highly connected nodes, so-called hubs. If the nodes with larger degree are in the majority, then the critical noise is larger and the minima disappear (not shown here).

To test theoretical predictions made with the mean-field theory in Gaussian approximation, we performed numerical simulations of random binary networks of stochastic rotators. The goal was to check whether the global oscillations can indeed be suppressed by increasing the network heterogeneity from a small to a moderate value. We took networks of $N = 10^4$ rotators and integrated Eq. (4.1) using the Heun method [66] with the time step of 0.5. Smaller time steps down to 0.05 had no significant effect on the simulation results.

We used the Viger-Latapy algorithm to generate the binary random networks [125]. In particular, we used the `C` library `igraph` [126] (version `0.6.5`), where the aforementioned algorithm is implemented.

Fig. 4.5 shows simulation results for two different networks. Both networks had the same average degree, but distinct variances. Compared with Fig. 4.4, the only difference in parameters of Fig. 4.5 is a larger value of the coupling strength which has no qualitative effect on global dynamics but simplifies the detection of the global oscillations. As can be seen from panels (a) and (b), an increase of the network heterogeneity leads to suppression of network's oscillations, in agreement with the theory. Panel (c) shows snippets of the phase distribution with the parameters from panel (a). As in Fig. 4.3(c)[upper left], we observe localized oscillations. One can recognize the distinct single-hump shape that is necessary for the Gaussian approximation.

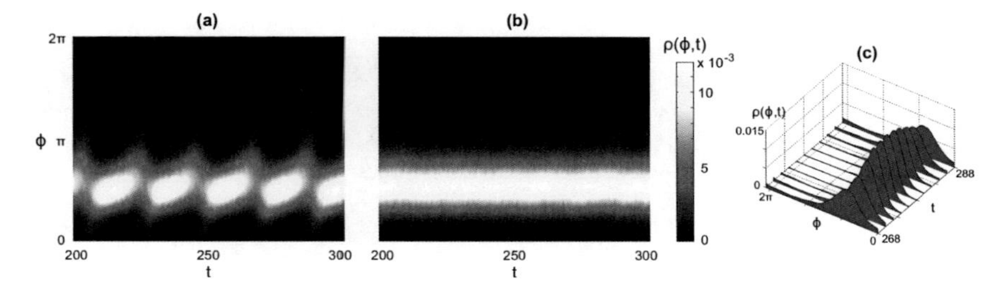

Figure 4.5.: Numerical simulations of random binary networks of $N = 10^4$ stochastic rotators. Shown are time-dependent probability densities of rotators phases, $\rho(\phi, t)$. In both panels the rotators are in the excitable regime with $a = 1.05$, $D = 0.095$ and $K = 2$. (a): Homogeneous binary network with parameters $\alpha_1 = 0.59$, $\alpha_2 = 0.39$ and $p = 0.05$ leading to the variance $\mathrm{Var}(\alpha)/\mathrm{VarMax} = 0.0076$. (b): Inhomogeneous binary network, $\alpha_1 = 0.97$, $\alpha_2 = 0.37$, $p = 0.05$ with the variance $\mathrm{Var}(\alpha)/\mathrm{VarMax} = 0.0684$. (c): Projections from the phase distribution in (a) for one period. From [122].

4.4. Cooperative behavior between self-oscillatory and excitable units

Collective dynamics in biological systems is in general a complex behavior that results from the interplay of non-identical, highly nonlinear and noisy elements [127]. Neuronal and cardiac rhythms for instance originate from interactions among pacemaker and excitable cells (see e.g. Refs. [8, 128–130] and [131–134], respectively). Motivated by these facts, we extend our previous work towards the investigation of the collective dynamics among coupled non-identical elements that can be either excitable or self-oscillatory. The latter shall model the pacemaking cells in neuronal or cardiac tissues, for instance. Furthermore, in order to make the model more general, individual coupling strengths are allowed to be different. Our setting enables us to study how certain correlations between the dynamics and the couplings on the microscopic level affect the macroscopic behavior of the system. Many works addressed the latter kind of question recently, see e.g. Refs. [74,76,78,81,135–147]. For interesting recent works that highlight the special interplay between dynamics and network structure in neuronal systems, we refer to [148, 149].

Of particular interest here are the works presented in [142, 145]. Zhang *et al.* considered Kuramoto oscillators coupled in a generalized complex network. Noteworthy, it was found that the crucial feature behind the emergence of explosive synchronization[1] is a positive correlation between the natural frequencies and the effective coupling strengths to the mean field [142]. Chen *et al.* studied effects of degree-frequency correlations in a population of FitzHugh-Nagumo neurons [145]. They extended in this way the finding of explosive synchronization to relaxation oscillators with two separated time scales.

The dynamical system that we study here puts emphasis on the phenomenon of excitability, both on the local and the global scale. Moreover, the coupling-frequency correlation considered in [142] shall motivate the specific formulation of our model. To this end, we investigate the noise-driven active rotator model introduced by Shinomoto and Kuramoto [112] with distributed natural frequencies and coupling strengths. Specifically, we analyze a system formed by two distinct parts of excitable and self-oscillating units, the first having subthreshold natural frequencies, while the other elements have frequencies above the excitation threshold.

We continue with a population of noise-driven active rotators [112], but differently than in the previous setting (4.1) we incorporate individual natural frequencies ω_i and coupling strengths K_i instead of a complex network A_{ij}:

$$\dot{\phi}_i = \omega_i - a \sin \phi_i + \frac{K_i}{N} \sum_{j=1}^{N} \sin(\phi_j - \phi_i) + \xi_i(t). \tag{4.14}$$

We will assume that the individual frequencies and coupling strengths are random numbers that are drawn from the same joint probability distribution $P(\omega, K)$, independently between the elements. It has to be emphasized that the values for the ω's and K's are chosen initially and then stay fixed during the whole evolution of the system. They represent frozen random variables ("quenched disorder"), which shall be given by some real numbers. We do not consider repulsive interactions here, that is the coupling strengths are non-negative.

For completeness, we write down the infinite chain of coupled complex-valued differential equations for the Fourier coefficients $\rho_n(t|\omega, K)$, which is almost the same as in Eq. (4.6). That is, for every pair (ω, K) we can write

$$\begin{aligned} \frac{\dot{\hat{\rho}}_n(t|\omega, K)}{n} &= \frac{a}{2}\Big[\hat{\rho}_{n-1}(t|\omega, K) - \hat{\rho}_{n+1}(t|\omega, K)\Big] - (Dn - i\omega)\hat{\rho}_n(t|\omega, K) \\ &+ \frac{K}{2}\left[\hat{\rho}_{n-1}(t|\omega, K)\langle\hat{\rho}_1(t|\omega', K')\rangle_{\omega,K} - \hat{\rho}_{n+1}(t|\omega, K)\langle\hat{\rho}_{-1}(t|\omega', K')\rangle_{\omega,K}\right]. \end{aligned} \tag{4.15}$$

While (4.15) provides an exact representation of the system, it is not possible to derive the solutions in an explicit way due to its hierarchical character. Since the Fourier coefficients rapidly decay with growing n, one can get accurate results by truncating the hierarchy at a large enough n. Here we use again the Gaussian approximation, Sec. 3.1, to derive

[1] Explosive synchronization was coined by the finding of a discontinuous synchronization transition in scale-free networks of Kuramoto oscillators with bistability between incoherence and partial synchronization [76].

an approximate dimensionality reduction that allows bifurcation analysis or even explicit solutions in important limiting cases. To this end, we seek a closure of the infinite set of equations (4.15) by assuming that in every subset of oscillators with the same individual quantities (ω, K), the distribution of the phases is Gaussian at all times with mean $m_{\omega,K}(t)$ and variance $\sigma^2_{\omega,K}(t)$ [99, 122].

For the first two cumulants of the Gaussian distributions, $\left\{m_{\omega,K}(t), \sigma^2_{\omega,K}(t)\right\}$ (4.8), we obtain now the following pair of ODE's:

$$\begin{aligned} \dot{m}_{\omega,K} &= \omega - \exp\left(-\sigma^2_{\omega,K}/2\right) \cosh \sigma^2_{\omega,K} \Big[a \sin m_{\omega,K} \\ &\quad - K\Big\langle \exp\left(-\sigma^2_{\omega',K'}/2\right) \sin\left(m_{\omega',K'} - m_{\omega,K}\right)\Big\rangle_{\omega,K}\Big], \\ \dot{\sigma}^2_{\omega,K}/2 &= D - \exp\left(-\sigma^2_{\omega,K}/2\right) \sinh \sigma^2_{\omega,K} \Big[a \cos m_{\omega,K} \\ &\quad + K\Big\langle \exp\left(-\sigma^2_{\omega',K'}/2\right) \cos\left(m_{\omega',K'} - m_{\omega,K}\right)\Big\rangle_{\omega,K}\Big]. \end{aligned} \tag{4.16}$$

Thus, for a continuous coupling-frequency distribution $P(\omega, K)$ the reduced system is still infinite-dimensional, because for any pair (ω, K) one has to solve the two differential equations (4.16), and all of those are coupled through the averages $\langle \ldots \rangle_{\omega,K}$. In order to obtain a low-dimensional system, we need to continue with a discrete coupling-frequency distribution $P(\omega, K)$ with a finite number of different ω's and K's. Indeed, interesting example systems are readily found, as shown in the next section.

Before coming to the integral part of our analysis, we would like to mention that one can also perform a variable transformation to the local mean-field variables (3.30):

$$\begin{aligned} \dot{r}_{\omega,K} &= -r_{\omega,K} D + \frac{1 - r^4_{\omega,K}}{2}\left[a \cos \Theta_{\omega,K} + K\Big\langle r_{\omega',K'} \cos\left(\Theta_{\omega',K'} - \Theta_{\omega,K}\right)\Big\rangle_{\omega,K}\right], \\ \dot{\Theta}_{\omega,K} &= \omega - \frac{r^{-1}_{\omega,K} + r^3_{\omega,K}}{2}\left[a \sin \Theta_{\omega,K} - K\Big\langle r_{\omega',K'} \sin\left(\Theta_{\omega',K'} - \Theta_{\omega,K}\right)\Big\rangle_{\omega,K}\right]. \end{aligned} \tag{4.17}$$

Recall that the mean phases are not defined in the case of vanishing mean-field amplitudes.

4.5. Bifurcation diagrams and order parameters for a concrete example

In the past, understanding the effects of heterogeneity benefited immensely by dividing the whole system into two subpopulations, see e.g. Refs. [122, 150–155]. Here we follow this strategy. On the basis of the reduced description derived in the previous section, we consider a mixed population consisting of two equally sized constituents; one half is chosen to be excitable and the other half shall be self-oscillating. This is realized by choosing one natural frequency below and the other one above the excitation threshold. Furthermore, both subpopulations shall have their own coupling strengths. Hence, for the coupling-frequency distribution we take a sum of two delta functions, $P(\omega, K) =$

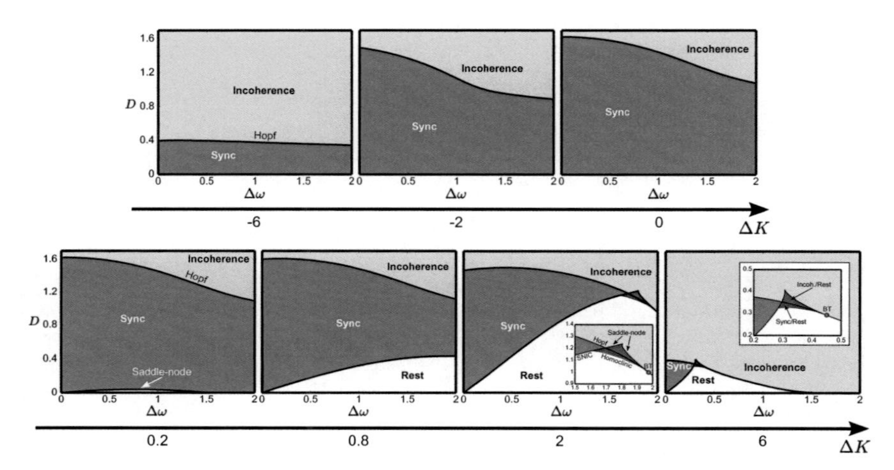

Figure 4.6.: Hopf and saddle-node bifurcations in the plane spanned by the noise intensity D and the frequency mismatch $\Delta\omega$, as obtained from a bifurcation analysis of the reduced system (4.18). Changes in the bifurcation diagram are shown as a function of the coupling mismatch ΔK, with a fixed average coupling strength of $K_0 = 4$. Green shaded areas with label "sync" represent the partially synchronized state, while in the yellow shaded regions no synchronized oscillations are found. In the white areas the system is at rest, in which the excitable units do not fire. For $\Delta K = 2$ and $\Delta K = 6$ insets show in more detail the parameter regions that correspond to bistable dynamics. The latter are found between the saddle-node bifurcation curves and the homoclinic bifurcation line which emanates from a Bogdanov-Takens bifurcation (BT). Moreover, the bistable region is separated into two parts by the Hopf bifurcation line. In the red area below the Hopf line partially synchronized and resting state coexist, while in the blue area resting and incoherence coexist. From [156].

$p\delta\left[(\omega, K)-(\omega_1, K_1)\right]+(1-p)\delta\left[(\omega, K)-(\omega_2, K_2)\right]$, with $p=0.5$, $0<\omega_1<1$, $\omega_2>1$, and $K_{1,2}>0$. In particular, we proceed with the following four-dimensional system [cf. Eq. (4.16)]:

$$\begin{aligned}\dot{m}_1 &= \omega_1 - \mathrm{e}^{-\sigma_1^2/2}\cosh\sigma_1^2\left[\sin m_1 + (K_1/2)\,\mathrm{e}^{-\sigma_2^2/2}\sin(m_1-m_2)\right],\\ \dot{\sigma}_1^2/2 &= D - \mathrm{e}^{-\sigma_1^2/2}\sinh\sigma_1^2\left\{\cos m_1 + (K_1/2)\left[\mathrm{e}^{-\sigma_1^2/2}+\mathrm{e}^{-\sigma_2^2/2}\cos(m_1-m_2)\right]\right\},\end{aligned} \tag{4.18}$$

where $\omega_1 = 1-\Delta\omega/2$ and $K_1 = K_0 - \Delta K/2$. The equations for $\dot{m}_2$ and $\dot{\sigma}_2^2$ are similar; just interchange 1's with 2's, and set $\omega_2 = 1+\Delta\omega/2$, $K_2 = K_0+\Delta K/2$. Indices $i=1,2$ are abbreviations for $\{\omega_i, K_i\}$. Henceforth, we call the differences in the natural frequencies and coupling strengths *frequency mismatch* ($\Delta\omega$) and *coupling mismatch* (ΔK), respectively. Note that the above choice is such that the average frequency and coupling strength are not affected by the mismatches. Equations for the four mean-field variables follow correspondingly from Eq. (4.17).

Similar problems were addressed in the context of oscillatory systems, where parts are inactivated due to aging [157–159]. Another mixed population of excitable and "driver" units was studied by Alonso and Mindlin [160]. Finally, the recent work [129] puts forward a detailed analysis of coupled theta neurons where both inherently spiking and excitable neurons are present.

From now on we study the collective behavior in system (4.18) with the help of `MATCONT` [123], namely in dependence on four dimensionless parameters: the noise intensity D, the frequency mismatch $\Delta\omega$, the coupling mismatch ΔK, and the average coupling strength K_0.

The frequency mismatch $\Delta\omega$ is varied in the interval $(0,2)$, restricting to positive natural frequencies. The coupling mismatch ΔK can take values between $(-2K_0, 2K_0)$. Positive (negative) values of ΔK can be referred to as positive (negative) coupling-frequency correlations, as long as there is a frequency mismatch $\Delta\omega>0$. Similar as in [108], we focus here on $K_0=4$.

Coupled excitable elements stay at rest, if they cannot globally surpass the excitation threshold. If they do, the question then is whether a macroscopic fraction of them fires in synchrony, which amounts to a partially synchronized state, or whether the firing is completely incoherent among the elements. Fig. 4.6 depicts the Hopf and saddle-node bifurcations that delineate those three states. Additional Hopf and saddle-node bifurcations that come after unstable equilibria can be neglected, because they do not affect the dynamics. As one would expect, increased frequency and coupling mismatches impede the emergence of collectively synchronized oscillations. Specifically, above the Hopf bifurcation line the oscillatory units fire incoherently, while below the Hopf line a synchronized firing sets in. Interestingly, a positive coupling-frequency correlation gives rise to a qualitative change in the global dynamics, since the saddle-node bifurcation shows up for $\Delta K>0$. Indeed the critical value for this phenomenon is found to equal $\Delta K_c=0$. This is visualized in Fig. 4.6 for a small coupling mismatch of $\Delta K=0.2$. We note that for $\Delta\omega\to 0$ the saddle-node bifurcation line always goes to vanishing noise intensity $D=0$, independently of ΔK. Below the saddle-node curve, the excitable elements are resting and do not fire. For increasing ΔK, the saddle-node line bends upwards, culminating

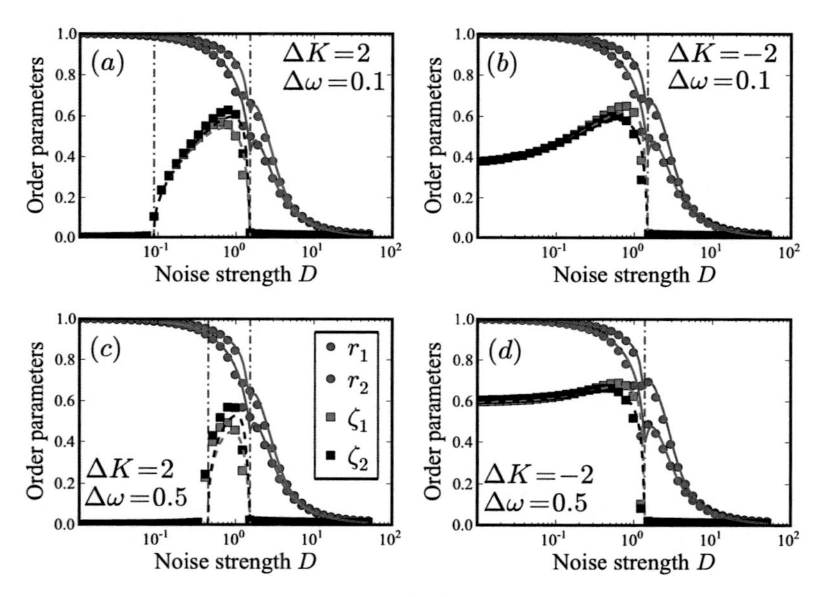

Figure 4.7.: Long-time averaged Kuramoto ($r_{1,2}$) and Kuramoto-Shinomoto order parameters [$\zeta_{1,2}$, cf. Eq. (4.19)], theory (lines) vs. simulation (dots). Vertical dash-dotted lines correspond to Hopf and SNIC bifurcations, respectively, as it can be extracted from Fig. 4.6. The average coupling strength is fixed at $K_0 = 4$. From [156].

in a Bogdanov-Takens bifurcation (BT), which is located at an intersection of the Hopf and the saddle-node lines. From the BT a homoclinic bifurcation line emanates, which ultimately merges with the saddle-node curve (then called a SNIC bifurcation line), see the insets in Fig. 4.6. We calculate the homoclinic bifurcations as follows. Starting at the BT we continue the Hopf bifurcation for some time steps, then switch the continuation to the limit cycle while tracking the period with the noise intensity D as the control parameter. At the homoclinic bifurcation the period of the limit cycle diverges. We accept the D values if they do not change anymore in the order of 10^{-4} upon approaching the divergence. The whole procedure is repeated until the homoclinic bifurcation line reaches the saddle-node curve. Both for $\Delta K = 2$ and $\Delta K = 6$ we save hereby in total eight pairs of $(\Delta\omega, D)$ and connect them by a line, see Fig. 4.6.

Importantly, the area between the saddle-node bifurcations and the homoclinic bifurcation line corresponds to bistable (hysteretic) dynamics. In particular, two qualitatively different bistable dynamics are separated by the Hopf bifurcation line; below it, the resting and the partially synchronized state coexist, whereas above there is a coexistence between two steady states, the resting and the incoherent state (compare with Ref. [99]).

Besides performing a bifurcation analysis, another way of characterizing the collective dynamics lies in calculating suitable order parameters. One of them is the classical Kuramoto order parameter, Eqs. (5.4), (5.5), which measures how similar the phase variables are to each other. However, it is not sufficient here to consider this order parameter, because in case of slowly varying phases, it would attain large values [112]. In the extreme case of resting elements, the Kuramoto order parameter would be even equal to unity, exactly as in the perfectly synchronized case. In order to distinguish between the resting and the synchronized state, one therefore needs to introduce an order parameter that decreases, if the elements collectively slow down. We consider here the well-known order parameter introduced by Kuramoto and Shinomoto [112]:

$$\zeta_{\omega,K}(t) = \left|\rho_1(t|\omega, K) - \overline{\rho_1(t|\omega, K)}\right|, \tag{4.19}$$

where $\rho_1(t|\omega, K) = r_{\omega,K}(t) \exp\left[i\Theta_{\omega,K}(t)\right]$ is the first coefficient of the Fourier series expansion of the one-oscillator probability density.

From now on, if we do not indicate an explicit time-dependence, we refer to long-time averages. In Fig. 4.7, we show the long-time averaged order parameters for certain sets of parameters, along with the bifurcation values as they can be extracted from Fig. 4.6. The three main regions mentioned for the bifurcation diagram 4.6 can be discriminated here as follows. While the Kuramoto order parameters are close to unity, and the Kuramoto-Shinomoto order parameters are nearly vanishing, the whole system is at rest, and single units do not fire. A partially synchronized oscillation on the global scale is achieved if both the Kuramoto-Shinomoto and the Kuramoto order parameters attain non-zero values. The third region is characterized by vanishing Kuramoto-Shinomoto and small Kuramoto order parameters. In this case, single units do fire, but in an incoherent way. In Fig. 4.7, panels (a) and (c), the humps in the Kuramoto-Shinomoto order parameters signal excitable behavior: for small noise intensities the population stays at rest, then at the SNIC bifurcation (first vertical dash-dotted line) one observes a transition to partial synchronization. Upon further increasing of the noise intensity the population becomes

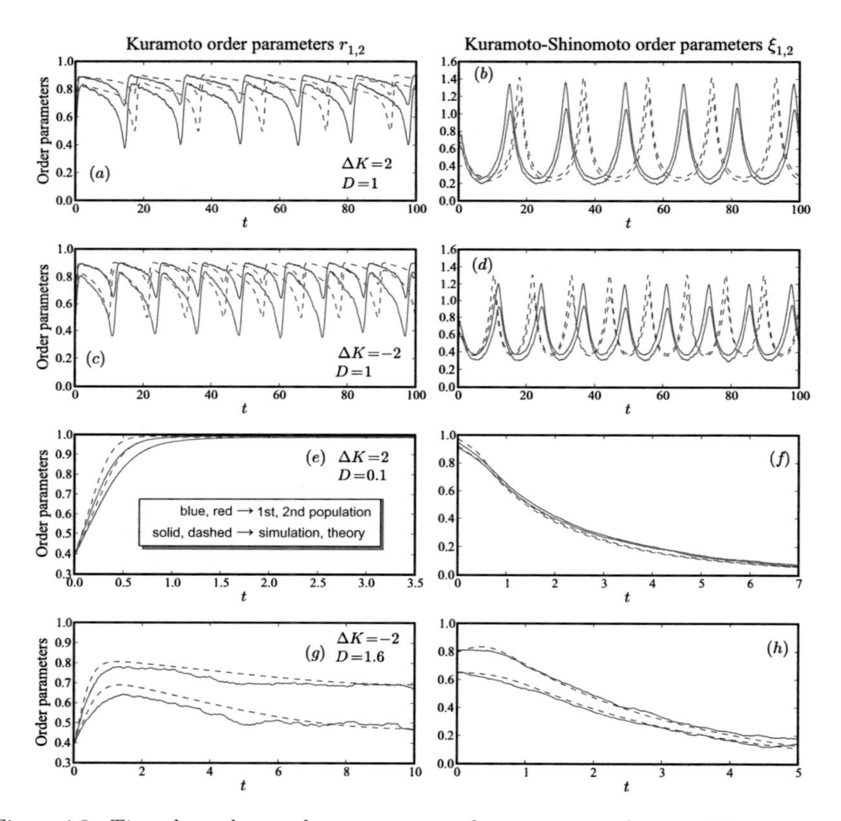

Figure 4.8.: Time-dependent order parameters, theory vs. simulation. The remaining parameters are frequency mismatch $\Delta\omega = 0.5$ and average coupling strength $K_0 = 4$. From [156].

completely incoherent, which happens precisely at the Hopf bifurcation (second vertical dash-dotted line). Panels (b) and (d) show no excitable behavior, but only a single transition at the Hopf bifurcation from partial synchronization to incoherence. Noteworthy however, the Kuramoto-Shinomoto order parameters depend non-monotonically on the noise intensity D, such that the highest level of synchronization is achieved at some non-zero noise intensity. The standard Kuramoto model cannot uncover this phenomenon.

Finally, one can observe that the theory agrees very well with the results from numerical simulations. We remark that the log scale is not necessary to appreciate the accuracy, the latter is chosen in order to emphasize the humps in the excitable regime (compare with Ref. [108]). The numerical simulations are conducted by integrating the stochastic equations of motion (4.14) using the Heun scheme with time step 0.05 and considering populations of $N = 10^4$ oscillators. Exactly one half of the population is assigned with frequencies and couplings $(\omega, K) = (\omega_1, K_1)$, and the second half with $(\omega, K) = (\omega_2, K_2)$. Initial conditions of the phases $\phi_i(t = 0)$ are Gaussian distributed with mean $m(t = 0) = 0$ and standard deviation $\sigma(t = 0) = \sqrt{2}$. Long-time averaged behavior of the order parameters is calculated by averaging the data between $t = [2500, 5000]$. For the theoretical lines we integrated the reduced system (4.18) with the same integration parameters.

Figure 4.8 shows that even the time-dependent behavior is correctly described by the reduced system (4.18). The parameters can be compared with the bifurcation diagram, Fig. 4.6. Note that the Kuramoto-Shinomoto order parameters can exceed unity as a function of time. Panels (a)–(d) reflect partially synchronized states, panels (e)–(f) represent resting behavior, and panels (g)–(h) correspond to incoherent dynamics. Apart from time shifts, the qualitative behavior is well predicted by the theory. In the collectively oscillating regime, the theoretical lines lag behind the simulation results for positive coupling mismatch ΔK, but the order is reversed for negative ΔK. Note that the initial values are not perfectly the same as a matter of fact. Finally, figure 4.8 illustrates a fundamental feature of the active rotator model, namely the inhomogeneous evolution of the phases. Such a property results in periodically oscillating mean-field amplitudes and order parameters under partial synchronization, see panels (a)-(d). Moreover, in the incoherent regime the classical Kuramoto order parameter does not vanish, see panel (g).

4.6. Summary and outlook

We have studied a well-known model for excitable dynamics, the so-called active rotators [112]. In the first part, besides investigating the effects of noise, we have put emphasis on the effects that emerge due to different coupling structures. To this end, we combined the network coarse-graining [cf. Sec. 2.1] with the Gaussian approximation [cf. Sec. 3.1] to reduce the dimensionality of the system in the thermodynamic limit. The order of the resulting system equalled the doubled number of different degrees occuring in the network. On the basis of the reduced system we performed numerical bifurcation analyses on regular and binary random networks. In this way we uncovered the parameter regions where the oscillator network shows synchronous or asynchronous behavior. Transitions between these regions were related to passages through Hopf and saddle-node bifurcations.

When the noise intensity was increased from zero, the saddle-node bifurcations marked the onset of global firing in the ensemble, whereas at the Hopf bifurcation the collective mode of synchronous oscillations disappeared. We were able to show that changing the network structure alone, without variations in the noise intensity and/or the excitation threshold, is sufficient to wander between the different collective states. This is interesting when keeping in mind that adaptations in the coupling structure are known to play a crucial role in the information processing among populations of neurons [9].

Specifically, deleting uniformly connections at all nodes in the network did not lead to new characteristics, since the coupling strength was then only rescaled by the fraction of connections. However, if nodes differed in their number of connections, an interesting observation was made in the excitable regime. We found that the critical noise intensity, at which the Hopf bifurcation occurs and small-scale oscillations are born, can possess a minimum at a moderate network heterogeneity. This was induced by the presence of a small fraction of highly connected nodes. For the functioning of neuronal networks, this might have meaningful implications, because such hub nodes naturally exist [9, 72]. It is well-known that neuronal networks are highly heterogeneous. Our analysis further suggested that this property implies the existence of large parameter regions to obtain collective spiking. If, however, a specific task requests a switching between different spiking patterns, reducing the heterogeneity in the connectivities becomes beneficial according to our results. The reason is that our findings suggest that only small noise variations are needed then to enable the switching. The optimum for such a task would then be given by a moderate coupling variance with a small fraction of hub nodes.

In the second part starting with Sec. 4.4, we have studied the active rotator model with distributed natural frequencies and coupling strengths in the globally coupled setting. Employing the same dimensionality reduction technique as before allowed us to analyze in detail a mixed population, where one half was chosen to be excitable, and the other half self-oscillatory. The distinction depended on whether the natural frequency was below or above the excitation threshold, respectively. Moreover, the elements of the two subpopulations differed in their individual coupling strengths. In this way we investigated how frequency and coupling mismatches affect the collective dynamics. In particular, we performed a numerical bifurcation analysis in the plane spanned by the noise intensity and the frequency mismatch and showed how these diagrams change as a function of the coupling mismatch. We found that both large frequency and coupling mismatches impede the emergence of synchronized oscillations. This is consistent with the expectation that oscillatory units which are more distinct, are harder to synchronize. More intriguingly however, we found that excitability in the whole system is only present, if the excitable elements have a weaker coupling than the self-oscillatory ones. In other words, a positive coupling-frequency correlation was necessary to cause the excitable behavior in the mixed population. We further found that bistability between various collective behaviors is only possible if the positive coupling-frequency correlation is strong enough. Such a phenomenon was previously reported only for systems without excitable dynamics, see e.g. Refs. [76, 142, 145]. For a discussion on the accuracy of the Gaussian approximation we refer to [156]. The embedded self-oscillatory units considered here can be regarded as pacemaker cells in neuronal [8] or cardiac [131, 133, 134] tissues. Our insights may provide a new perspective on the emergence of excitable behavior on the global scale, as it is observed e.g. in nonlinear optical cavities [161]. It would be interesting to further analyze effects of asymmetries in the natural frequencies and the coupling strengths, as it was done e.g. in [160] for a deterministic system.

5. Noisy oscillators with asymmetric attractive-repulsive interactions

In the previous chapters we have explored the transition from incoherence to partial synchronization in various settings. We have seen that with the incorporation of excitability, a qualitatively new collective state is possible, where the oscillators are mostly at rest. What other collective states can emerge, and under which conditions? The effects of asymmetric interactions remain particularly elusive, even more so if negative coupling strengths are allowed. We would like to shed light on this topic in the current chapter.

Motivated by Daido's seminal work on "oscillator glasses" [162], Hong and Strogatz obtained interesting results recently in this direction. In a series of papers [153, 154, 163], they investigated two mutually globally coupled populations of Kuramoto phase oscillators that differed in their coupling strengths. Two scenarios of mixed attractive and repulsive interactions were distinguished (positive and negative coupling strengths, respectively). In the first case some oscillators' phases repel the phases of all the others, while the remaining attract all the other phases [163]. This situation resembles neuronal networks with excitatory and inhibitory connections [164, 165]. In the second case some oscillators tend to align with the mean field, while others oppose it favoring an antiphase alignment [153, 154]. This version is analogous to sociophysical models of opinion formation [166]. Surprisingly, only the second scenario led to enriched dynamics beyond the traditional order-disorder transition.

Here, we unify both coupling scenarios by considering noisy phase oscillators that have the same natural frequency, but different in- and out-coupling strengths. Specifically, we consider a stochastic version of the Kuramoto model with twofold disordered coupling strengths:

$$\dot{\phi}_i(t) = \omega_0 + \frac{K_i}{N} \sum_{j=1}^{N} G_j \sin(\phi_j - \phi_i) + \xi_i(t), \tag{5.1}$$

Using the notion of give-and-take as a metaphor, oscillator i contributes to the mean field with weight G_i and at the same time incorporates the mean activity with weight K_i into its own dynamics. Accordingly, we call K_i in- and G_i out-coupling strength, respectively. Grouping together oscillators with the same coupling strengths, the number of different pairs (K_i, G_i) coincides with the number of subpopulations.

We consider two equally-sized subpopulations denoted by "1" and "2." Hence, the oscillators are distinguished by a pair of coupling strengths, (K_1, G_1) or (K_2, G_2), and all of those can be positive or negative. We choose the parametrization

$$K_{1,2} = K_0 \pm \frac{\Delta K}{2}, \quad G_{1,2} = G_0 \pm \frac{\Delta G}{2}. \tag{5.2}$$

K_0 and G_0 are average in- and out-coupling strengths, while ΔK and ΔG give corresponding mismatches. If $|\Delta K|/2 > |K_0|$ or $|\Delta G|/2 > |G_0|$, then half of the couplings are positive (attractive) and half negative (repulsive). In such cases we speak of mixed interactions. Note that Eq. (5.2) leads to point symmetries, because changing $(K_0, G_0) \to (-K_0, -G_0)$ or $(\Delta K, \Delta G) \to (-\Delta K, -\Delta G)$ yields the same situations. In the following, all oscillators have the same constant natural frequency ω_0. Therefore, by virtue of the rotational symmetry, we can set $\omega_0 = 0$ without loss of generality. Time-dependent disorder $\xi_i(t)$ is included as Gaussian white noise, $\langle \xi_i(t) \rangle = 0$, $\langle \xi_i(t)\xi_j(t') \rangle = 2D\delta_{ij}\delta(t-t')$. The angular brackets denote averages over different realizations of the noise and the single non-negative parameter D denotes the noise intensity. The noise terms $\xi_i(t)$ can be regarded as an aggregation of various stochastic processes [100]. In Refs. [153, 154, 163] it is found for the deterministic case $D = 0$ that mixed out-couplings alone do not enable more than partial synchronization, whereas mixed in-couplings yield traveling waves reached through diametrically synchronized states.

Since our system has no spatial coordinate, the traveling waves mentioned here are observed only through the aggregate of the phase evolutions and differ from the spatiotemporal patterns usually investigated, for instance in Refs. [167–171].

Here, we intermingle the two types of mixed attractive-repulsive interactions mentioned before and we explore in particular whether traveling waves persist in the presence of noise $D > 0$. Without loss of generality all subsequent results are obtained with $D = 0.5$, but for illustration we keep D in the derivations.

"Discordant synchronization" is used here as an umbrella term for situations where the ensemble splits into two partially synchronized clusters. This will include traveling waves and π-states, the latter being the extreme form of discordance with two oscillator populations anti-aligned to each other.

We investigate the thermodynamic limit $N \to \infty$, where propagation of molecular chaos [52] allows us to describe each population by a *one-oscillator* probability density $\rho_{1,2}(\phi, t) \equiv \rho(\phi, t|K_{1,2}, G_{1,2})$. Normalization requires $\int_0^{2\pi} \rho_{1,2}(\phi, t) d\phi' = 1 \ \forall \ t$. For given coupling strengths $K_{1,2}$ and $G_{1,2}$, $\rho_{1,2}(\phi, t)d\phi$ denotes the fraction of oscillators with phase between ϕ and $\phi + d\phi$ at time t. The densities are governed by the nonlinear Fokker-Planck equations [54, 80]:

$$\frac{\partial \rho_{1,2}}{\partial t} = D\frac{\partial^2 \rho_{1,2}}{\partial \phi^2} - \frac{\partial}{\partial \phi}\left[K_{1,2} R \sin(\Theta - \phi)\, \rho_{1,2}\right] . \tag{5.3}$$

The global mean-field amplitude $R(t)$ and phase $\Theta(t)$ follow from a superposition:

$$R(t)\mathrm{e}^{i\Theta(t)} = \frac{1}{2}\left[r_1(t)\ G_1\ \mathrm{e}^{i\Theta_1(t)} + r_2(t)\ G_2\ \mathrm{e}^{i\Theta_2(t)}\right], \tag{5.4}$$

Note that subpopulations of different sizes can be treated simply by rescaling $G_{1,2}$. The local mean-field variables obey

$$r_{1,2}(t)\mathrm{e}^{i\Theta_{1,2}(t)} = \int_0^{2\pi} d\phi' \mathrm{e}^{i\phi'} \rho_{1,2}(\phi', t) . \tag{5.5}$$

The level of synchrony in the two subpopulations is measured separately by $r_{1,2}(t)$, whereas for the global measure we take the classical Kuramoto order parameter $r(t) \equiv$

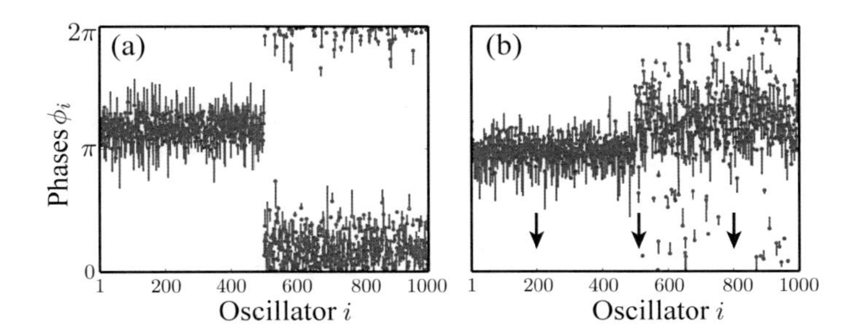

Figure 5.1.: Snapshots of oscillators' phases. (a) Stationary π-state at $K_0 = 1$, mean phase difference $\delta = \pi$, no net flux; (b) Traveling wave at $K_0 = 3$, flow downwards with stationary profile, indicated by black arrows; red dots depict phase values, blue lines of different lengths and directions indicate instantaneous frequencies. Remaining parameters: $\Delta K = 3$, $G_0 = 2$, $\Delta G = 10$, $D = 0.5$. For visualization the two subpopulations are separated into two halves. From [172].

$\frac{1}{2}\left|r_1(t)\,\mathrm{e}^{i\Theta_1(t)} + r_2(t)\,\mathrm{e}^{i\Theta_2(t)}\right|$, which differs from Eq. (5.4) by taking out $G_{1,2}$. The order parameter values lie between zero (incoherence) and 1 (complete synchronization). The variables $\Theta(t)$ and $\Theta_{1,2}(t)$ stand for the corresponding mean phases. Of special interest is the phase lag $\delta(t)$, i.e. the difference in the mean phases of the two populations, $\delta(t) = \Theta_1(t) - \Theta_2(t)$.

In the sequel, variables without a dot or other indicated time dependence refer to the long-time limit. Let us outline the four qualitatively different self-organized states observed here after some transient dynamics. (i) In the incoherent state the whole population of oscillators rotates asynchronously, $r_{1,2} = 0$. (ii) The classical partially synchronized state has zero phase lag, $r_{1,2} > 0$, $\delta = 0$. We use "zero-lag sync" as a shortcut to denote this state. (iii) The π-state describes a partially synchronized state, where the two subpopulations are anti-aligned to each other, $r_{1,2} > 0$, $\delta = \pi$. (iv) In the traveling wave state the whole population is also partially synchronized, but oscillates with a frequency different from the frequency of single oscillators. This spontaneous change in rhythm is induced by a phase lag that is neither zero nor π, $r_{1,2} > 0$, $0 < \delta < \pi$. We calculate the wave speed as

$$\Omega = \frac{1}{N}\sum_{i=1}^{N}\langle\dot{\phi}_i\rangle_t, \tag{5.6}$$

where $\langle\ldots\rangle_t$ represents a long-time average [153, 154].

In Fig. 5.1 example snapshots from simulations of $N = 10^5$ oscillators are shown ($N = 10^3$ are equidistantly chosen for visualization). Figure 5.1(a) displays a π-state with $r_1 = 0.98$, $r_2 = 0.89$, $r = 0.05$, $|\delta| = \pi$ and $|\Omega| = 0$. Figure 5.1(b) shows a traveling wave state with $r_1 = 0.98$, $r_2 = 0.75$, $r = 0.80$, $|\delta| = 1.42$ and $|\Omega| = 3.17$. It is equally possible that the wave runs in the other direction, depending on initial phases and realization of the noise. Note that perfect synchrony, $r_{1,2} = 1$, cannot be achieved with finite coupling strengths, if an infinitesimal amount of noise is present.

5.1. Bifurcation diagrams

In order to analytically investigate the collective dynamics that are governed by (5.3)-(5.5), we approximate the phase distributions in the two populations by time-dependent Gaussians, cf. Sec. 3.1. As a result, we obtain a reduced three-dimensional system of coupled ODE's:

$$\begin{aligned}
\dot{r}_1 &= -r_1 D + \frac{1-r_1^4}{4} K_1 \left[r_1 G_1 + r_2 G_2 \cos\delta \right], \\
\dot{r}_2 &= -r_2 D + \frac{1-r_2^4}{4} K_2 \left[r_2 G_2 + r_1 G_1 \cos\delta \right], \\
\dot{\delta} &= -\frac{\sin\delta}{4} \left[\left(r_1^{-1} + r_1^3\right) K_1 r_2 G_2 + \left(r_2^{-1} + r_2^3\right) K_2 r_1 G_1 \right].
\end{aligned} \tag{5.7}$$

All the four aforementioned collective states are fixed points of (5.7) with $\dot{r}_{1,2} = \dot{\delta} = 0$. Two types of fixed point solutions have to be distinguished, because there are two possibilities that $\dot{\delta} = 0$ holds:

$$\delta = m\pi, \; m \in \mathbb{Z}, \tag{5.8}$$

$$0 = \left(r_1^{-1} + r_1^3\right) K_1 r_2 G_2 + \left(r_2^{-1} + r_2^3\right) K_2 r_1 G_1. \tag{5.9}$$

Equation (5.8) describes zero-lag and π-states, whereas Eq. (5.9) underlies traveling waves. Intermediate phase lags $0 < \delta < \pi$ cause spontaneous drifts, because according to Eqs. (5.7) and (5.9) the common frequency of the traveling waves obeys

$$\lim_{t\to\infty} \dot{\Theta}_1 = \lim_{t\to\infty} \dot{\Theta}_2 = \sin\delta \, \frac{r_2^{-1} + r_2^3}{4} \, K_2 G_1 r_1. \tag{5.10}$$

Two more equations are obtained from imposing $\dot{r}_{1,2} = 0$ in (5.7):

$$\begin{aligned}
r_1 &= \frac{r_2}{G_1 \cos\delta} \left[\frac{4D}{\left(1 - r_2^4\right) K_2} - G_2 \right], \\
r_2 &= \frac{r_1}{G_2 \cos\delta} \left[\frac{4D}{\left(1 - r_1^4\right) K_1} - G_1 \right].
\end{aligned} \tag{5.11}$$

With (5.8), (5.9) and (5.11) we have three coupled equations for three unknowns, $r_{1,2}$ and δ. No stationary solution with $\delta = \pi/2$ can be found, but the singularities $G_{2,1} = 0$ and $K_{1,2} = 0$ turn out to have a special meaning. In particular, if one of the in-coupling strengths $K_{1,2}$ vanishes, the corresponding population remains incoherent. Numerical

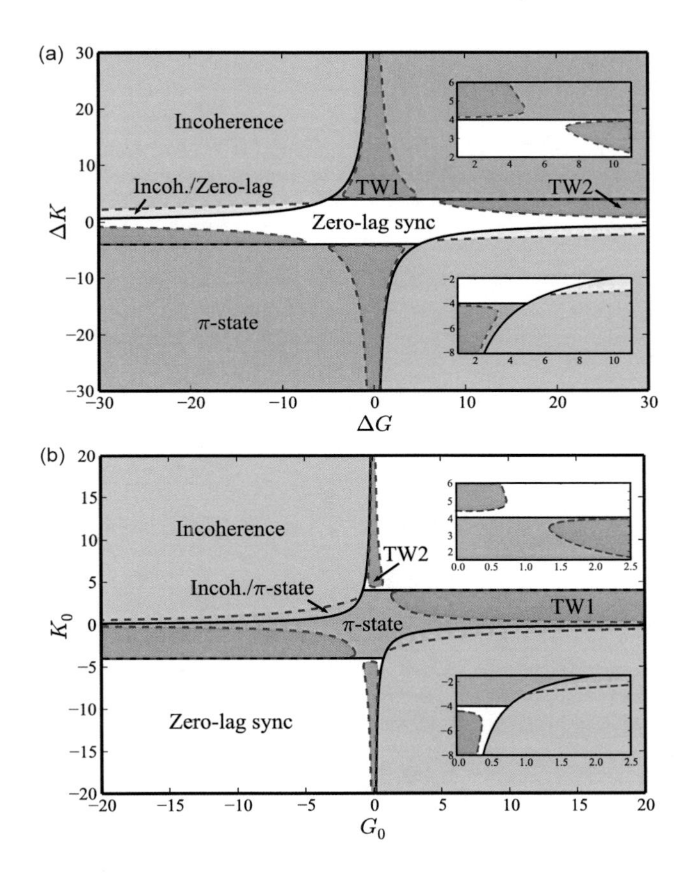

Figure 5.2.: Bifurcation diagrams for (a) $K_0 = 2$, $G_0 = 3$ and (b) $\Delta K = 8$, $\Delta G = 2$. Solid lines from Eqs. (5.12), (5.13), dashed lines with MATCONT [123]. "Zero-lag sync" denotes partially synchronous states with zero phase lag between the two subpopulations. Parameter regions with lag $\delta = \pi$ labeled as "π-states". Traveling waves (TW1 and TW2) have non-zero wave speed Ω. Insets show enlarged areas. Modified from [172].

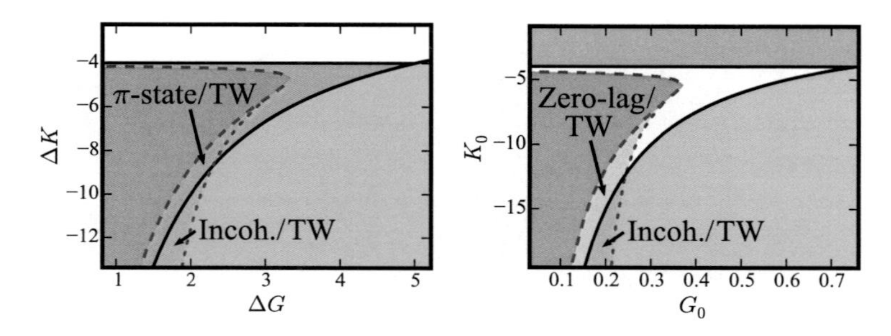

Figure 5.3.: This figure is supplementary to Fig. 5.2 by showing further zooms. Coexistence of traveling waves with other collective states is found. Red dotted lines obtained with MATCONT, plotted only here. Modified from [172].

continuation around this point shows that in order to avoid a negative local order parameter, $r_{1,2} < 0$, which is unphysical, the whole population transfers to a π-state. In our parametrization this first critical condition can be written as

$$\Delta K_{c_1} = \pm 2K_0. \tag{5.12}$$

One can show that in general the incoherent state, $r_{1,2} \equiv 0$, loses linear stability, if the noise intensity falls below a certain value [cf. Eq. (5.16)]. This happens at

$$(\Delta K \Delta G)_{c_2} = 8D - 4K_0G_0. \tag{5.13}$$

Finally, as outlined below, the intersection given by $G_{2,1} = 0$ and (5.13) coincides with the origin of bistability. Note that the aforementioned conditions are exact. Figures 5.2(a) and 5.2(b) depict bifurcation diagrams in the planes spanned by the coupling mismatches (ΔK, ΔG) and the average coupling strengths (K_0, G_0), respectively. Solid lines are given by the critical conditions (5.12) and (5.13). Dashed lines are obtained on the basis of the reduced system (5.7) with the help of MATCONT [123]. We detect branch and limit points, since at all lines one eigenvalue vanishes, except at the ones given by (5.12), because those do not correspond to real bifurcations, but delineate two analogous partially synchronous states: zero-lag and π-states. We additionally test all these findings by numerically calculating the eigenvalues of the Jacobian of (5.7) with the fixed points given by (5.8)-(5.11). When the two lines given by (5.12) and (5.13) intersect, the boundaries (5.12) cease to exist. We emphasize two distinct routes to TW states; TW1 is surrounded by π-states, TW2 by classical zero-lag sync states. Delimiting lines approach each other, see insets for enlarged areas. We further find bistability between incoherence and zero-lag or π-states, see panels (a) and (b) in Fig. 5.2, respectively. The bistable areas are circumscribed by two lines that intersect at the points given by $\Delta G = \pm 2G_0$ and (5.13). The location of this intersection determines the type of bistability. Interestingly, traveling and non-traveling wave states can coexist in small parameter regions. We show

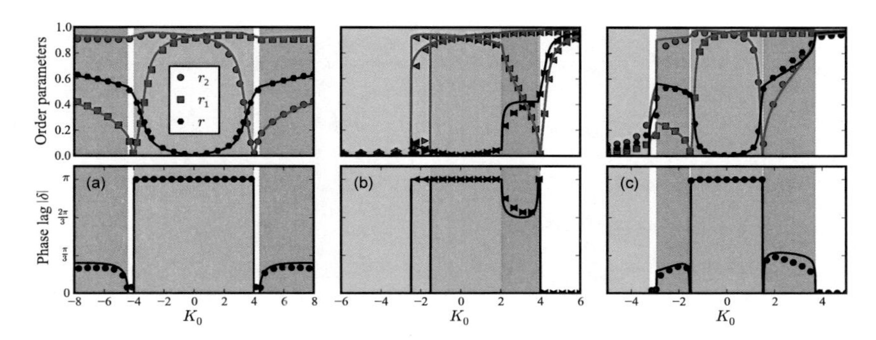

Figure 5.4.: Order parameters r_2, r_1, r and phase lag $|\delta|$ as a function of average in-coupling strength K_0. (a) $G_0 = 0$, (b) $G_0 = 2$, remaining parameters in both cases: $\Delta K = 8$, $\Delta G = 2$. (c) $\Delta K = 3$, $\Delta G = 10$, $G_0 = 2$ [cf. Eq. (5.2)]. In all panels, dots are obtained by integrating the full system (5.1) with $N = 10^4$ oscillators. All lines are obtained by numerically solving Eqs. (5.8)-(5.11). In order to unfold the hysteresis in (b), besides forward continuation ($\triangleright$) as in the other cases, a backward continuation ($\triangleleft$) is performed. Use of colored regions as in Fig. 5.2. From [172].

this in Fig. 5.3. In particular it is observed that traveling waves can coexist with complete incoherence, as well as with π-states and zero-lag partially synchronous states.

Figure 5.2 suggests certain conditions for the various collective states. In order to observe π-states, mixed attractive-repulsive in-couplings have to be included. Traveling waves surrounded by π-states are possible, if one includes mixed in-couplings without mixed out-couplings. In contrast, traveling waves surrounded by zero-lag sync states exist, if there are mixed out-couplings and a non-zero mismatch without mixing in the in-couplings. Bistability between incoherence and zero-lag sync requires lack of mixed in-couplings, but the presence of mixed out-couplings. Finally, bistability between incoherence and π-states is possible by combining mixed in-couplings with vanishing mixing in the out-couplings. These conditions appear to supplement consistently the observations in Refs. [153, 154, 163]. In particular, we verified numerically that the traveling waves surrounded by zero-lag sync can also be observed in the setting studied in Ref. [163], if a small mismatch in the in-couplings is present, as suggested by our bifurcation diagram in Fig. 5.2(a).

It is worth asking how crucial asymmetric interactions are for the discordant synchronization patterns discussed here. As it is easily seen from Eq. (5.2), the interactions are symmetric if in- and out-coupling strengths balance each other such that the equation $0 = K_0\Delta G - G_0\Delta K$ holds. This condition can be projected onto straight cuts through the parameter space. The demarcations would not cross the traveling wave areas [see Fig. 5.2(a) and 5.2(b)]; instead they would divide the bifurcation diagrams into parameter regions that contain both routes to traveling waves, i.e. TW1 and TW2. In other words, asymmetric interactions are needed to get traveling waves. Interestingly however, those straight cuts would go through the π-state regimes. Thus, asymmetry in the interactions is not a necessary ingredient to observe π-states. This conclusion is not evident from the works presented in [153, 154, 163].

5.2. Simulation results

Additional insights can be gained by numerically solving the three coupled equations (5.8)-(5.11) to get $r_{1,2}$ and δ. The only subtlety is that one has to factor in the bifurcation values previously obtained in order to correctly choose between (5.8) and (5.9). Solutions are shown in Fig. 5.4, and compared with the results from numerical simulations. For the latter, initial phases are randomly chosen from the uniform distribution $[-\pi, \pi]$. For each K_0 value a long-time average is taken over $t \in [2500, 5000]$ with integration time step $dt = 0.01$. Upper panels depict the order parameters, while in the lower panels the corresponding phase lags are shown. The colored regions match those in Fig. 5.2 and discriminate the different collective states. In Fig. 5.4(b) one can see that at the transition from incoherence to π-state the suborder parameters abruptly jump from zero to high values in a hysteretic manner. For very long time averages it is expected that the hysteresis is washed out due to noise-induced jumps between the two stable steady states. We do not report this here. In Fig. 5.4(c), around $K_0 \approx -3$, the abrupt change in the phase lag and the non-vanishing order parameters signal extended stability of traveling waves, as discussed for Fig. 5.2(c). Such bistable dynamics appears to be a promising topic for future studies. In general, the abrupt changes and the local minima in the order parameters as a function of the average in-coupling strength K_0, as presented in Fig. 5.4, are of vital interest on their own, see e.g. Refs. [76, 173] and [174], respectively.

In Fig. 5.5 we compare the common frequency obtained from the reduced system, Eq. (5.10), with numerical simulations. It serves as an alternative measure to the wave speed (5.6), which is calculated from the individual instantaneous frequencies that do not exist in the analytical treatment. Therefore for the wave speed no comparison with theory is being made. In Fig. 5.5 zero-lag synchronous and π-states become unstable in the regime of traveling waves. One can observe that both measures, the common frequency and the wave speed, highlight equally well the onset of traveling waves. As mentioned before, the waves emerge in frequency pairs, meaning that they can travel in both directions, depending on realization of random numbers.

In Fig. 5.6 results of numerical simulations are superimposed on smaller theoretical bifurcation diagrams. In Figs. 5.6(a)-5.6(c) local and global order parameters are depicted, while in Fig. 5.6(d) the wave speed is plotted. For each of the 100×100 data points in the $K_0 \times G_0$ grid, the equations of motion (5.1) are integrated with $N = 10^4$ oscillators and observables are then averaged over time, $t \in [100, 500]$. Different initial conditions are chosen in Figs. 5.6(e)-5.6(g) in order to find bistability in numerical simulations (50×50 data points there). Specifically, in Figs. 5.6(e)-5.6(g) the area circumscribed by dashed and solid lines shows π-states: the local order parameters $r_{1,2}$ attain large values, but

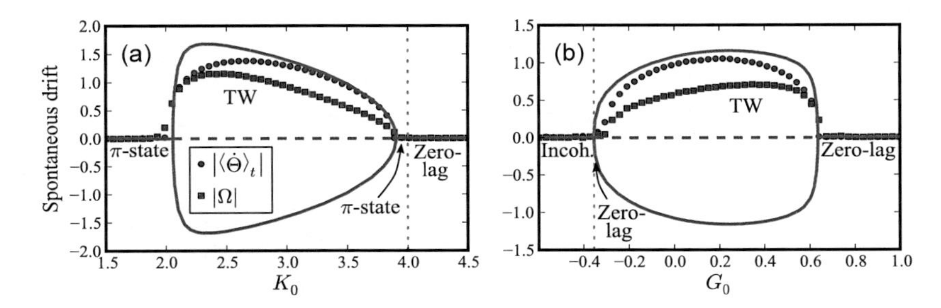

Figure 5.5.: The spontaneous drift measured by common frequency $\left|\langle\dot{\Theta}\rangle_t\right|$, Eq. (5.10) (theory [lines] vs. simulation [dots]) and wave speed $|\Omega|$, Eq. (5.6) (only simulation). (a) $G_0 = 2$, (b) $K_0 = 6$; remaining parameters: $\Delta K = 8$, $\Delta G = 2$ [cf. Eq. (5.2)]. Simulation of $N = 10^4$ oscillators, averaging over $t \in [200, 500]$, step size $dt = 0.01$. Modified from [172].

due to the anti-phase alignment given by the phase lag π, the total order r is small. In Figs. 5.6(a)-5.6(c) this area is filled with incoherence. Hence, numerical simulations agree again very well with theoretical results. We remark that bistability between incoherence and zero-lag synchronous states, as predicted by the theory, can be analogously found by varying the initial conditions (not shown).

5.3. Towards an arbitrary number of populations

Here we discuss the transition from incoherence to partial synchrony in an arbitrary number of interacting populations of arbitrary sizes. To this end, oscillators with the same pair of in- and out-coupling strengths are again grouped into one population. Then the nonlinear Fokker-Planck equation for the one-oscillator probability density $\rho(\phi, t|K, G)$ reads

$$\frac{\partial \rho(\phi, t|K, G)}{\partial t} = D\frac{\partial^2 \rho(\phi, t|K, G)}{\partial \phi^2} - \frac{\partial}{\partial \phi}\left[\left\{KR(t)\sin\left[\Theta(t) - \phi\right]\rho(\phi, t|K, G)\right\}\right]. \tag{5.14}$$

The global mean-field amplitude $R(t)$ and phase $\Theta(t)$ are given by

$$R(t)\mathrm{e}^{i\Theta(t)} = \langle r_{K',G'}(t)\; G'\; \mathrm{e}^{i\Theta_{K',G'}(t)}\rangle_{K,G}. \tag{5.15}$$

The averages $\langle \ldots \rangle_{K,G} \equiv \int dK' \int dG' \ldots P(K', G')$ take into account all in- and out-coupling strengths, K and G, via the corresponding joint probability distribution $P(K, G)$. In Eq. (5.15) averages are taken over the local mean-field variables, $r_{K,G}(t)\mathrm{e}^{i\Theta_{K,G}(t)} = \int_0^{2\pi} d\phi' \mathrm{e}^{i\phi'} \rho\left(\phi', t|K, G\right)$. By a linear stability analysis of the incoherent state, $\rho(\phi, t|K, G) = 1/(2\pi)\ \forall K, G, t$, one can exactly derive the critical noise intensity for the synchronization

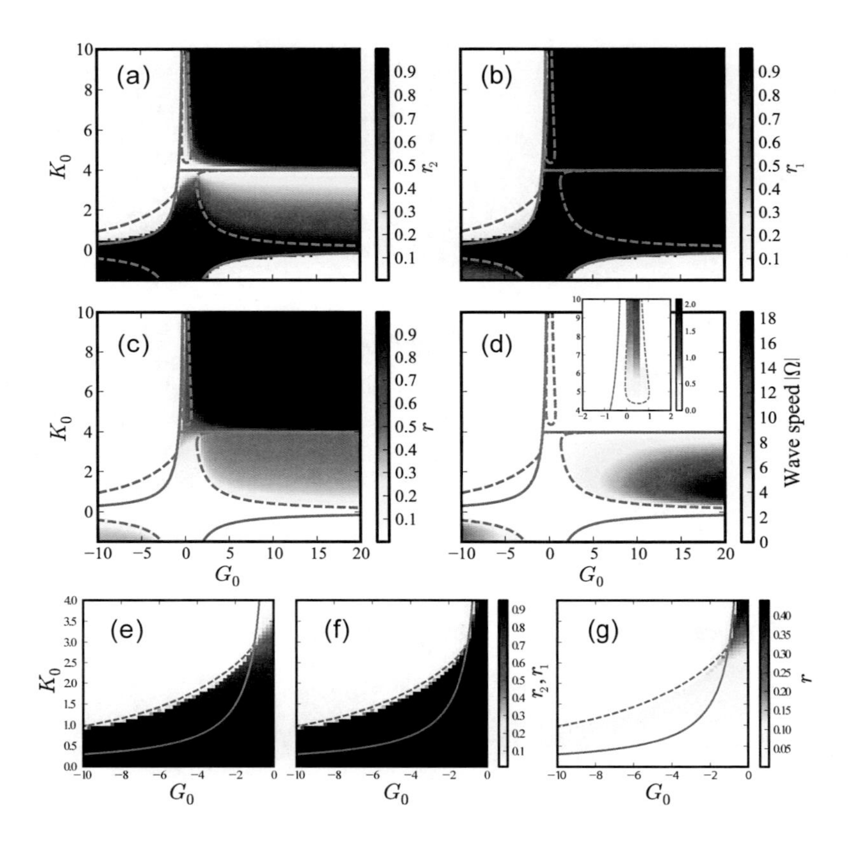

Figure 5.6.: Simulations (colormaps) vs. theory (lines); panels (a) and (b) show local order parameters r_2 and r_1, while in (c) the global order r is depicted. Panel (d) depicts the absolute value of the wave speed $|\Omega|$. Inset shows a zoom-in. Simulation performed with initial phases randomly chosen from uniform distribution $[-\pi, \pi]$. In panels (e)-(g) simulations are repeated with initial phases $\phi_i(0) = 1 \; \forall i$. Remaining parameters are the same as in Fig. 5.2 (b). From [172].

transition:

$$D_c = \frac{\langle K'G' \rangle_{K,G}}{2}. \tag{5.16}$$

This result follows from adapting the derivation [54, 80] to the present case. Above D_c the whole ensemble is incoherent, below D_c the incoherent state loses stability and partial synchrony is observed. Now we give an alternative derivation on the basis of the Gaussian approximation. It recovers the exact condition (5.16) and yields an instructive generalization of the equations in the main part of the manuscript. Inserting the Gaussian approximation for all phase distributions $\rho(\phi, t|K, G)$ into Eqs. (5.14)-(5.15), we obtain a set of coupled differential equations for the local mean-field amplitudes $r_{K,G}(t)$ and mean phases $\Theta_{K,G}(t)$:

$$\begin{aligned} \dot{r}_{K,G} &= -r_{K,G}D + \frac{1 - r_{K,G}^4}{2} K \left\langle r_{K',G'} G' \cos\left(\Theta_{K',G'} - \Theta_{K,G}\right) \right\rangle_{K,G}, \\ \dot{\Theta}_{K,G} &= \frac{r_{K,G}^{-1} + r_{K,G}^3}{2} K \left\langle r_{K',G'} G' \sin\left(\Theta_{K',G'} - \Theta_{K,G}\right) \right\rangle_{K,G}. \end{aligned} \tag{5.17}$$

Let us consider small perturbations $\delta r_{K,G}(t)$ of the incoherent state, $r_{K,G}(t) = 0\ \forall K, G, t$. The perturbations may give rise to zero-lag synchronous or π-states, $\Theta_{K',G'} - \Theta_{K,G} = m\pi$, $m = 0, 1$. Accordingly, we can separate the network of networks into two groups, 1 and 2, which contain subpopulations with coupling strength pairs (K, G) that lead to the same mean phases $\Theta_{K,G}$. At the same time, the mean phases of the subpopulations in the groups 1 and 2 differ by π. Linearizing around the perturbations $\delta r_{K,G}(t)$ in the two groups separately, we obtain from Eqs. (5.17):

$$\begin{aligned} \left[\dot{\delta r}_{K,G}\right]_1 &= -\left[\delta r_{K,G}\right]_1 D + \frac{1}{2} K_1 \left\{ \left[\left\langle \delta r_{K',G'} G' \right\rangle_{K,G}\right]_1 - \left[\left\langle \delta r_{K',G'} G' \right\rangle_{K,G}\right]_2 \right\}, \\ \left[\dot{\delta r}_{K,G}\right]_2 &= -\left[\delta r_{K,G}\right]_2 D - \frac{1}{2} K_2 \left\{ \left[\left\langle \delta r_{K',G'} G' \right\rangle_{K,G}\right]_1 - \left[\left\langle \delta r_{K',G'} G' \right\rangle_{K,G}\right]_2 \right\}. \end{aligned} \tag{5.18}$$

Here, the two separated networks of networks are labeled by the indices 1 and 2. According to Eq. (5.15) the two perturbations can be put together as $\delta R(t) C = \left[\left\langle \delta r_{K',G'} G' \right\rangle_{K,G}\right]_1 - \left[\left\langle \delta r_{K',G'} G' \right\rangle\right]_2$, where C is some constant coming from an arbitrary global mean phase. As a result we obtain

$$\dot{\delta R}(t) = \left[-D + \frac{1}{2} \langle K'G' \rangle_{K,G}\right] \delta R(t), \tag{5.19}$$

which leads to the critical condition (5.16).

5.4. Summary and outlook

We have explored the rich dynamics that emerge from asymmetric in- and out-coupling strengths among two mutually globally coupled oscillator populations. We considered the illustrative example of identical noisy Kuramoto phase oscillators with non-uniform and mixed attractive-repulsive interactions. The two populations were observed to partially synchronize either in-phase or with a constant phase lag to each other. We referred to the latter as "discordant synchronization." The phase lags induced spontaneous drifts. As a result, traveling waves were formed in which the whole population oscillated with a different frequency than the individual units. However, in the state of maximal discordance, when the two partially synchronized populations are anti-aligned to each other, the spontaneous drifts disappeared. Correspondingly, we revealed two distinct routes to traveling waves, one through diametral two-cluster states, the other one through classical one-cluster states. Since the latter are ubiquitously investigated in the literature, we expect the second route to traveling waves to be more prevalent in general. Appropriate experimental setups made up of two constituents are conceivable, realized e.g. with laser systems [175, 176], metronomes [177], electric circuits [178] or chemical Belousov-Zhabotinsky oscillators [179, 180].

The Gaussian approximation (cf. Sec. 3.1) led to a three-dimensional system of coupled ODE's. This reduced system allowed a thorough bifurcation analysis and further analytical treatment, in excellent agreement with numerical simulations of a large but finite number of oscillators. We found that physically relevant singularities constitute a significant part of the bifurcation scenario. We further showed which collective states can coexist. Our results help to understand the emergence of discrepancies between individual and collective rhythms, as is observed especially in neuronal networks [164, 165]. If the connection strengths were capable to slowly vary in time, one could expect temporal patterns reminiscent of the high-frequency rhythmic events observed in hippocampal networks [181, 182]. Attractive (positive) and repulsive (negative) couplings are often associated with excitatory and inhibitory connections among neurons. This is reasonable, since positive couplings tend to increase synchrony, which is also the case for excitatory connections in the brain, while negative and inhibitory connections both tend to decrease synchrony [183]. Exceptions however exist, see e.g. Ref. [184]. The individual out-coupling strengths considered here are particularly suitable to emulate the role of excitatory or inhibitory neurons. This was pointed out in Ref. [163] by referring to Dale's principle, according to which a neuron releases the same set of neurotransmitters at all its synapses [185]. Such a comparison would become even more applicable by including an excitation threshold into the system, as in the last chapter 4. Whether the combination of mixed attractive-repulsive interactions on the level of in- and out-coupling strengths is experimentally relevant, remains an interesting topic for the future.

6. Synchronization in the stochastic Kuramoto-Sakaguchi model

When oscillatory units are coupled together, their collective frequency may differ from the average of the natural frequencies. We have seen in Sec. 5 that such an additional drift can be induced by phase lags between subpopulations of oscillators. Instead of studying the emergence of phase lags due to negative coupling strengths, we consider here a sinusoidal coupling function that explicitly contains a phase lag parameter α. This model is also known as the Kuramoto-Sakaguchi model [94,186]. Notably, this extended system allows concrete applications, see in particular the works on Josephson arrays [187,188]. We study a stochastic version of it, where temporal fluctuations serve as the only source of disorder:

$$\dot{\phi}_i(t) = \omega_0 + \frac{K}{N}\sum_{j=1}^{N}\sin(\phi_j - \phi_i + \alpha) + \xi_i(t). \tag{6.1}$$

By virtue of the rotational symmetry in the model, we can set $\omega_0 = 0$ without loss of generality. Again the time-dependent disorder is modeled by Gaussian white noise with noise intensity D. Since here we are interested in purely attractive coupling, we focus on phase lags that obey $|\alpha| \leq \pi/2$.

6.1. Application of Gaussian approximation

We try to get analytical insights by following the steps of the Gaussian approximation, Sec. 3.1. Therefore, let us first write down the nonlinear Fokker-Planck equation (FPE) that governs the evolution of the one-oscillator probability density $\rho(\phi, t)$ in the thermodynamic limit $N \to \infty$:

$$\frac{\partial \rho}{\partial t} = D\frac{\partial^2 \rho}{\partial \phi^2} - \frac{\partial}{\partial \phi}\left[Kr\sin\left(\Theta - \phi + \alpha\right)\rho\right] , \tag{6.2}$$

with the classical Kuramoto order parameters $r(t)$ and $\Theta(t)$, cf. Eq. (1.3). After inserting the Fourier series for the density $\rho(\phi, t)$ into the FPE, we find an infinite chain of coupled differential equations for the Fourier coefficients [compare with Eq. (3.7)]:

$$\frac{\dot{\hat{\rho}}_n}{n} = \frac{K}{2}\left(\hat{\rho}_{n-1}\hat{\rho}_1 \mathrm{e}^{i\alpha} - \hat{\rho}_{n+1}\hat{\rho}_{-1}\mathrm{e}^{-i\alpha}\right) - Dn\hat{\rho}_n, \tag{6.3}$$

with $n = 1, 2, \ldots, \infty$. Note that the equations are complex-valued with $\hat{\rho}_0 = 1$ and $\hat{\rho}_{-n} = \hat{\rho}_n^*$.

Assuming a wrapped Gaussian distribution for $\rho(\phi, t)$, the whole population is fully characterized by the first two moments of that time-dependent Gaussian. As a result, an explicit closure in the hierarchy (6.3) is achieved, since all higher Fourier coefficients can be expressed through the first one [99]. In terms of the mean-field amplitude $r^g(t)$ and mean phase $\theta^g(t)$, $\hat{\rho}_1^g(t) = r^g(t)\exp\left[i\theta^g(t)\right]$, we find the following two-dimensional system of differential equations [again we use the superscript g to denote expressions obtained within the Gaussian theory, compare with Eq. (3.14)]:

$$\begin{aligned}\dot{r}^g &= \frac{r^g}{2}\left\{K\cos\alpha\left[1-\left(r^g\right)^4\right]-2D\right\},\\ \dot{\theta}^g &= \frac{K}{2}\sin\alpha\left[1+\left(r^g\right)^4\right].\end{aligned} \tag{6.4}$$

Observe that r^g couples into θ^g, but not the other way around, and therefore a full solution can be readily obtained.

6.2. Connection between common frequency and synchronization

We derived the solution for the Kuramoto order parameter r^g already in Sec. 3.2 for the case without phase lag, $\alpha = 0$. Taking that solution, one only has to substitute K by $K\cos\alpha$. Then we obtain

$$r^g(t) = \left|\frac{K-K_c}{\left(\frac{K-K_c}{\left[r^g(0)\right]^4}-K\right)\mathrm{e}^{-2t\cos\alpha(K-K_c)}+K}\right|^{1/4}. \tag{6.5}$$

The critical coupling strength is now given by

$$K_c = \frac{2D}{\cos\alpha}. \tag{6.6}$$

One can show that this value is exact, in the sense that the incoherent solution indeed becomes linearly unstable beyond K_c [54,150,189]. Phase lags therefore increase the critical coupling strength and hence hinder synchronization; Eq. (6.5) additionally suggests an interesting side remark, namely that a phase lag slows down the temporal evolution of the order parameter. The solution for the mean phase immediately follows from (6.4):

$$\theta^g(t) = \theta^g(0) + \frac{Kt\sin\alpha}{2}\left[1+\left(r^g\right)^4(t)\right], \tag{6.7}$$

where $r^g(t)$ is the solution (6.5). In the long-time limit the Kuramoto order parameter equals the one derived before, cf. Eq. (3.22), only with the different critical coupling strength (6.6):

$$r^g = \left[1-\left(K_c/K\right)\right]^{1/4}. \tag{6.8}$$

In Fig. 6.1(a) this order parameter is plotted as a function of K/K_c for different phase lags α. The theoretical results collapse on one line. As a consequence, the Gaussian approximation becomes less accurate upon increasing the phase lag α. This can be expected,

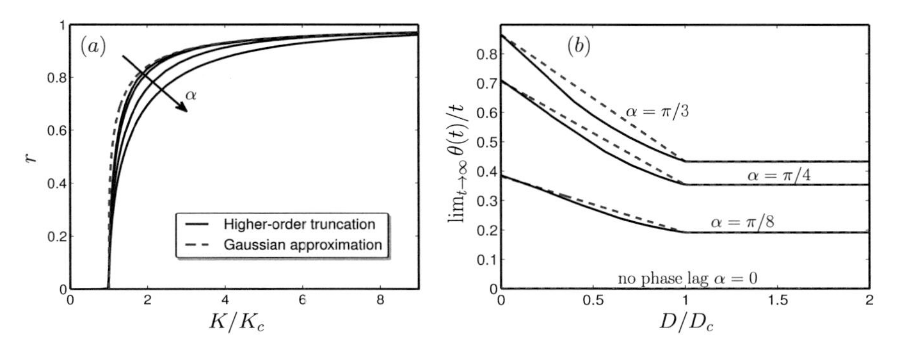

Figure 6.1.: Comparison between theoretical results obtained with the Gaussian approximation (blue dashed lines) and the practically exact results obtained by evaluating numerically Eq. (6.3) with truncation at $n = 60$ (black solid lines). In panel (a) the Kuramoto order parameter is shown, Eq. (6.8), for different phase lags α that are also used in panel (b) to depict the common frequency, Eq. (6.9)

because the phase distribution becomes skewed, if a phase lag exists, but the Gaussian distribution is symmetric by definition.

Using (6.8) in (6.7), the common frequency in the long-time limit is found to obey

$$\lim_{t\to\infty} \frac{\theta^g(t)}{t} = \begin{cases} (K/2)\sin\alpha, & K \leq K_c, \\ (K\sin\alpha - D\tan\alpha), & K > K_c. \end{cases} \tag{6.9}$$

The completely incoherent regime $K \leq K_c$ is not meaningful, because the mean phase $\theta(t)$ is not defined there. But in the ordered regime $K > K_c$, we see from Eq. (6.9) that the population collectively oscillates with a common frequency that increases with the level of synchronization. The highest frequency is obtained for vanishing noise intensity $D = 0$, when the population is perfectly synchronized. The absence of any phase lag constitutes here the only scenario, in which the common frequency remains zero in the co-rotating frame, which means that the common frequency then equals the natural frequency ω_0. This all is shown in Fig. 6.1(b).

Remarkably, a related connection between the firing rate and the output correlation has been reported for the spiking activity of cortical neurons [190].

6.3. Towards a new phenomenological theory

Here we present a heuristic reasoning that leads to an improved theory. In the first step we consider the regime close to the critical coupling strength K_c (6.6), where the higher Fourier coefficients are expected to play a minor role, see Fig. 3.2. From (6.3) we obtain

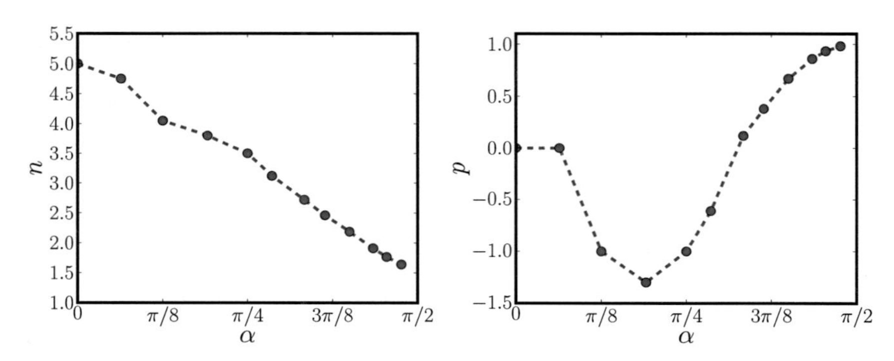

Figure 6.2.: The two fitting parameters $n(\alpha)$ and $p(\alpha)$ as a function of the phase lag α. They appear in Eqs. (6.14), (6.15).

the first three Fourier coefficients,

$$\begin{aligned}\dot{\hat{\rho}}_1 &= \frac{K}{2}\left(\hat{\rho}_1 \mathrm{e}^{i\alpha} - \hat{\rho}_2\hat{\rho}_1^* \mathrm{e}^{-i\alpha}\right) - D\hat{\rho}_1,\\ \dot{\hat{\rho}}_2 &= K\left(\hat{\rho}_1^2 \mathrm{e}^{i\alpha} - \hat{\rho}_3\hat{\rho}_1^* \mathrm{e}^{-i\alpha}\right) - 4D\hat{\rho}_2,\\ \dot{\hat{\rho}}_3 &= \frac{3K}{2}\left(\hat{\rho}_2\hat{\rho}_1 \mathrm{e}^{i\alpha} - \hat{\rho}_4\hat{\rho}_1^* \mathrm{e}^{-i\alpha}\right) - 9D\hat{\rho}_3.\end{aligned} \tag{6.10}$$

Now, slightly above K_c, as an approximation we can truncate (6.10) via $\hat{\rho}_3 \approx 0$ and $\dot{\hat{\rho}}_2 \approx 0$ (Ref. [2], p. 286). This yields a Stuart-Landau equation:

$$\dot{\hat{\rho}}_1 = \left(\frac{K}{2}\mathrm{e}^{i\alpha} - D\right)\hat{\rho}_1 - \frac{K^2}{8D}\left|\hat{\rho}_1\right|^2\hat{\rho}_1. \tag{6.11}$$

Separating into amplitude and phase, $\hat{\rho}_1 = r(t)\exp[-i\theta(t)]$, we obtain with (6.6) and (6.11):

$$\begin{aligned}\dot{r} &= \frac{(K-K_c)\cos\alpha}{2}r - \frac{K^2}{4K_c\cos\alpha}r^3,\\ \dot{\theta} &= -\frac{K}{2}\sin\alpha.\end{aligned} \tag{6.12}$$

The equation for the mean phase gives no useful information, in particular it does not depend on the order parameter $r(t)$ and therefore cannot recover the connection between common frequency and synchronization, which was resolved with the help of the Gaussian approximation [compare Eqs. (6.7), (6.9) and Fig. 6.1]. This could be expected, because Eq. (6.12) holds only asymptotically close to K_c, where the order parameter tends to vanish. The upshot however is that we can now derive the correct scaling of the order parameter, where the phase lag α shows up not only in the critical coupling K_c. That is, in the long-time limit it follows from (6.12):

$$r = \frac{\sqrt{2K_c}\cos\alpha}{K}\sqrt{K-K_c}. \tag{6.13}$$

We attempt to combine phenomenologically the differential equations for the order parameters from (6.4) and (6.12) with the help of two fitting parameters that depend on the phase lag α. We call them $n(\alpha)$ and $p(\alpha)$. Specifically, we include these two parameters in the following form:

$$\dot{r} = \frac{r}{2}\cos\alpha\left[K\left(1-r^4\right)-K_c\right]-\frac{r^3}{4}\left(\frac{K_c}{K}\right)^{n(\alpha)}\frac{K^2}{K_c}\left[p(\alpha)\cos\alpha+\frac{1-p(\alpha)}{\cos^2\alpha}\right]. \tag{6.14}$$

The stationary solution in the long-time limit $t\to\infty$ reads

$$\begin{aligned} r &= \frac{1}{2}\sqrt{\frac{\sqrt{64K(K-K_c)K_c^2\cos^4\alpha+\Upsilon^2}-\Upsilon}{2KK_c\cos^2\alpha}}, \\ \Upsilon &\equiv K^2\left(\frac{K_c}{K}\right)^n\left[2-p+p\cos(2\alpha)\right]. \end{aligned} \tag{6.15}$$

We would like to mention that similar (arbitrary) fitting procedures can be found for related problems, see [191] and the appendix E in [192]. For several phase lag values α we search for the $n(\alpha)$ and $p(\alpha)$ values that provide a good fit between theory, Eq. (6.15), and numerical truncation of (6.3). The values that we find thereby are collected in Fig. 6.2. Note that for $\alpha = 0$ the parameter p has no effect in (6.14). Similarly for $\alpha = 0.2$ we find that the p parameter can be varied in a large interval around zero without revealing a significant change in the order parameter r. For this reason we set $p = 0$ for those two α values, see Fig. 6.2.

In Fig. 6.3 we compare the new result (6.15) with the scaling result close to criticality (6.13), the result from the Gaussian approximation (6.8) and the practically exact result obtained by numerical truncation of (6.3) at $n = 60$. The four panels present a subset of the phase lag values α that we checked, cf. Fig. 6.2. In all cases we find that two α-dependent fitting parameters allow to obtain very accurate stationary order parameters with the new phenomenological theory for all coupling strengths K and phase lags α.

6.4. Summary and outlook

We have investigated the Kuramoto-Sakaguchi model, which differs from the traditional Kuramoto model by imposed phase lags in the sinusoidal coupling function. In particular, we considered a stochastic version where the only source of disorder comes from temporal fluctuations acting on the instantaneous frequencies. First, we applied the Gaussian approximation (cf. Sec. 3.1) to obtain a two-dimensional reduced system. We were able to solve this system. Thereby we recovered the exact critical coupling strength for the onset of synchronization. Additionally, the Gaussian approximation correctly predicted the appearance of drifts in the partially synchronous state, if phase lags are present. Similarly as in chapter 5, this leads to a common frequency that differs from the natural frequency of single oscillators. Inaccuracies were observed for larger phase lags that result from increasingly asymmetric phase distributions. In order to cover this fact, we extended the Gaussian theory in a phenomenological way by combining it with the scaling result

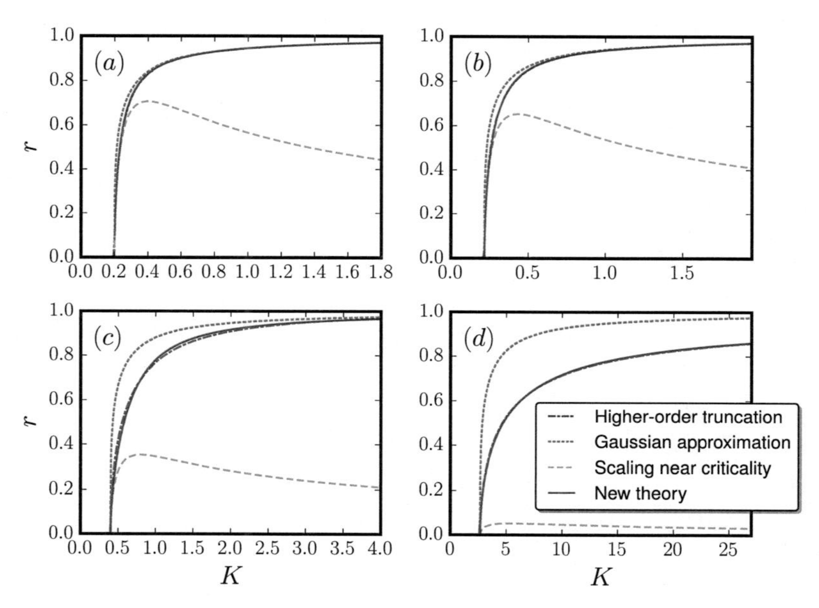

Figure 6.3.: Comparison of the different theoretical results [Gaussian approximation (6.8), critical scaling (6.13) and new phenomenological theory (6.15)] vs. numerical truncation of (6.3) at $n = 60$ which corresponds to the practically exact solution. The four panels depict results for different phase lags: (a) $\alpha = 0$, (b) $\alpha = \pi/3$, (c) $\alpha = \pi/8$, and (d) $\alpha = \pi/2.1$.

that holds near the critical coupling strength. This extended theory has two fitting parameters that depend on the phase lag. In this framework we obtained very accurate results for all phase lag values.

In a next step it would be interesting to use the extended theory to investigate the effects of noise on the collective states among identical oscillators, in which a coexistence of coherence and incoherence can be observed, the so-called chimera states [193, 194]. A minimal model for this phenomenon was investigated by Abrams *et al.* which consists of two equally sized populations that are all-to-all coupled within and between each other [195]. The couplings are symmetric and the same within the populations, but the links connecting the two populations are supposed to have a smaller strength. A noisy version would read

$$\dot{\phi}_i^\sigma(t) = \xi_i^\sigma(t) + \sum_{\sigma'=1}^{2} \frac{K_{\sigma\sigma'}}{N_{\sigma'}} \sum_{j=1}^{N_{\sigma'}} \sin(\phi_j^{\sigma'} - \phi_i^\sigma + \alpha), \tag{6.16}$$

where the index $\sigma = 1, 2$ labels the two subpopulations and $N_1 = N_2$, $K_{12} = K_{21} < K_{11} = K_{22}$. First results for (6.16) were obtained in [196] by numerically truncating the Fourier mode representation, but the investigation based on evolution equations for the order parameters is missing. For this purpose we expect an ansatz such as in Eq. (6.14) to prove useful. Since chimera states were recently observed in experiments (see e.g. [177, 179]), which are inherently affected by noise, it is of crucial interest to better understand the effects of noise on these intriguing coherence-incoherence patterns.

7. Conclusions

The goal in this thesis was to shed new light on the effects of noise and network structure on the collective dynamics in Kuramoto-type phase oscillator models. Two methods turned out to be particularly useful thereby: (1) coarse-graining the networks by grouping together oscillators with the same individual quantities, (2) assuming that the oscillators' phases are governed by time-dependent Gaussian distributions. The first method enables mean-field descriptions that can be used to study the critical conditions for certain collective states, while the second method allows to find low-dimensional systems that can be used for comprehensive bifurcation analyses and further analytical treatments. Both methods were used separately before, the network coarse-graining for instance in [40–42], and for Gaussian dimensionality reductions see e.g. [98, 99]. The novel contribution of our work was to combine the two into one fruitful method. The emerging picture is the one of interacting time-dependent Gaussian distributions, where each Gaussian represents one subpopulation which is built from the phases of the oscillators that share the same individual quantities. Here, we used this method in various interesting example systems.

In Chap. 2 we have investigated noisy Kuramoto phase oscillators in dense small-world networks. Considering the infinite system size limit, we used a mean-field approach via network coarse-graining to derive a nonlinear Fokker-Planck equation for the one-oscillator probability density. From there we derived the critical coupling strength for the onset of synchronization. We began with the case where natural frequencies and degrees do not depend on each other. The critical coupling strength was then found to be given by two factors, unifying previous results as follows. One factor described the deterioration of synchrony by noise and frequency mismatch, as it was derived in [54, 80]. The second factor was given by the ratio of the first two moments of the degree distribution, hence showing how synchronization is enhanced by increasing either the number of connections or the variance thereof. This result was previously obtained in [41, 42]. We compared the theoretical results with numerical simulations via finite-size scaling analysis and found good agreement. We discussed the discrepancies and showed that the mean-field approach breaks down in very sparsely connected networks, in particular if random shortcuts are absent. Next we considered an example where at each node the dispersion of the natural frequencies is correlated with the degree, i.e. each oscillator draws its frequency from a given distribution which has a degree-dependent width. The main finding was that if the correlation is positive and of certain power, the critical coupling strength for the onset of synchronization becomes independent of the heterogeneity in the number of connections. Without noise this masking effect happened for a linear correlation, irrespective of the specific shape of the frequency distribution (assumed to be unimodal and symmetric), but with noise the correlation had to be stronger and tuned to the frequency distribution to obtain the same critical coupling in two networks of different degree heterogeneity.

In Chap. 3 we have laid down the foundations of the Gaussian approximation, where one makes the ansatz that the phases of the oscillators are governed by a time-dependent Gaussian distribution. We used this approximation in the Kuramoto model with noise as the only source of disorder to find a reduced description, which consisted of two uncoupled ordinary first-order differential equations. As a consequence, we could easily find a full solution. Specifically, we derived an expression for the time-dependent order parameter revealing approximately the level of synchronization at any point in time and for any coupling strength. Noteworthy, the well-known exact critical coupling was reproduced by the Gaussian theory. We further showed that the theory yields very accurate order parameters in the incoherent and sufficiently synchronized regime. While the Gaussian theory did not reproduce the square-root scaling in the vicinity of the critical coupling, it appeared to give an improved explicit upper bound for the stationary order parameter. Finally, we presented how the Gaussian approximation can be combined with a network coarse-graining to obtain an extended theory that includes quenched disorder such as distributed frequencies or degrees. We demonstrated the potential of this approach by recovering analytically in a simple way the previously obtained synchronization transition point in complex networks.

In Chap. 4 we have investigated a well-known model for excitable dynamics, the so-called active rotators [112]. The effects of noise in excitable systems were comprehensively studied before [22]. Much less known are the combined effects of noise and network structure. As a first step, we considered noise-driven active rotators in complex networks. Again we combined the Gaussian approximation with the coarse-graining valid for uncorrelated random networks. Through this mean-field approach we arrived at a lower-dimensional set of integro-differential equations. The order of this system equaled two times the number of different degrees occuring in the network. We performed numerical bifurcation analyses on regular and binary random networks, when the system was two- and four-dimensional, respectively. Deleting uniformly connections of all nodes resulted effectively only in a rescaling of the coupling strength, but if nodes differed in their degrees, an interesting observation was made in the excitable regime. We found that the critical noise intensity, at which the Hopf bifurcation occurs and small-scale oscillations are born, can possess a minimum at a moderate network heterogeneity. This stood in sharp contrast to what we have seen in self-oscillatory systems, where the critical noise intensity grew (almost always) linearly with the variance in the degrees. The new finding was induced by the presence of a small fraction of highly connected nodes. Next we studied the active rotator model with distributed natural frequencies and coupling strengths in an all-to-all coupled network. We used a similar low-dimensional system as before to analyze a mixed population, where one half was chosen to be excitable, the other half self-oscillatory. The distinction depended on whether the natural frequency was below or above the excitation threshold, respectively. Moreover, the elements of the two subpopulations differed in their coupling strengths. Numerical bifurcation analysis revealed that both frequency and coupling mismatch impede the emergence of synchronized oscillations. Intriguingly however, we found that excitability in the whole system is only present, if the excitable elements have a smaller coupling strength than the self-oscillatory ones. This may provide a new perspective on the emergence of excitable behavior on the global scale.

In Chap. 5 we have explored the rich dynamics that emerge from asymmetric in- and out-coupling strengths among two mutually globally coupled oscillator populations. As an illustrative example we considered identical noisy Kuramoto phase oscillators with non-uniform and mixed attractive-repulsive interactions. With the Gaussian approximation we derived a three-dimensional system of coupled ODE's. This reduced system allowed a thorough bifurcation analysis and further analytical treatment. We observed that the two populations can partially synchronize in-phase or with a constant phase lag to each other, which induced spontaneous drifts. As a result, traveling waves were formed in which the whole population oscillated with a different frequency than the individual units, as it is also observed in neuronal networks [164, 165]. However, when the two partially synchronized populations were anti-aligned to each other, the spontaneous drifts disappeared. Correspondingly, we revealed two distinct routes to traveling waves, one through diametral two-cluster states, the other one through classical one-cluster states. Since the latter are ubiquitously investigated in the literature, we expect the second route to traveling waves to be more prevalent in general. Attractive (positive) and repulsive (negative) couplings are often associated with excitatory and inhibitory connections among neurons. This is reasonable, since positive couplings tend to increase synchrony, which is also the case for excitatory connections in the brain. In contrast, negative and inhibitory connections have in common that they tend to decrease synchrony [183]. Exceptions however exist, see e.g. Ref. [184]. The individual out-coupling strengths considered here are particularly suitable to emulate the role of excitatory or inhibitory neurons. This was pointed out in Ref. [163] by referring to Dale's principle, according to which a neuron releases the same set of neurotransmitters at all its synapses [185]. Such a comparison would become even more applicable by including an excitation threshold as in Chap. 4.

In Chap. 6 we have investigated the Kuramoto-Sakaguchi model, which differs from the traditional Kuramoto model by phase lags explicitly imposed in the sinusoidal coupling function. We considered a stochastic version where the only source of disorder comes from temporal fluctuations acting on the instantaneous frequencies. We were able to solve the reduced system obtained from the Gaussian approximation. Thereby we recovered the exact critical coupling strength for the onset of synchronization. We observed the appearance of drifts in the partially synchronous state, if phase lags are present. Similarly as in chapter 5, this led to a common frequency that differs from the natural frequency of single oscillators. Inaccuracies between theoretical and numerical results were observed for larger phase lags. They came from increasingly asymmetric phase distributions. In order to cover this fact, we extended the Gaussian theory in a phenomenological way by combining it with the scaling result that holds near the critical coupling strength. Very accurate results were then obtained for all phase lag values. It would be interesting to use the extended theory to investigate the effects of noise on the so-called chimera states [193,194] (coexistence of coherence and incoherence among otherwise identical oscillators), since they can be observed in a variant of the Kuramoto-Sakaguchi model [195]. Chimera states were argued to have a loose connection to unihemispheric sleep of e.g. some sea mammals, in the sense that the awake side of the brain shows desynchronized electrical activity, whereas the sleeping side remains highly synchronized [195].

In summary, we have studied large networks (from regular to heterogeneous random ones) of noise-perturbed phase oscillators with additional time-independent disorder in the natural frequencies and coupling strengths. We also incorporated excitable dynamics through an excitation threshold. In a population of excitable and self-oscillatory elements, the latter could be considered as the pacemaker cells. Moreover, we included mixed positive and negative coupling strengths, and discussed the relationship to excitatory and inhibitory connections. In such a setup of mixed couplings, phase lags can emerge that cause the common frequeny of the whole population to be different from the single oscillator frequency. We also indicated that directly imposing phase lags yields similar traveling wave patterns. Noteworthy, phase lags can model the effects of small time delays. Taking together the population dynamics that we have obtained, and the methods that we have developed thereby, we believe that our work contributes to a better understanding of the collective dynamics observed in neuronal networks [197], in particular with respect to the emergence of discrepancies between individual and collective rhythms. The main aspects missing here concern the spiking dynamics, adaptation mechanisms, synaptic plasticity as well as modular and hierarchical coupling structures. It is our hope that the results contained in this thesis will fruitfully inform more sophisticated models of neuronal network dynamics and the functioning therein.

A. Directed networks

Real-world systems generically constitute directed networks, i.e. complex networks where the connections are equipped with directions, so that nodes are not necessarily mutually coupled. In neuronal networks the gap junction networks are mostly considered to be undirected (or bidirectional), but the networks formed chemical synapses are evidently directional, see for instance [72]. The consequence of directionality is that the corresponding adjacency matrix becomes asymmetric with elements $A_{ij} = 1$, if unit i couples to unit j, and otherwise $A_{ij} = 0$. Complex topologies of real-world networks can be encoded into the adjacency matrix, and decoded by counting all the in- and out-degrees, which are given by

$$k_j^{\text{in}} = \sum_{i=1}^{N} A_{ij}, \quad k_i^{\text{out}} = \sum_{j=1}^{N} A_{ij}. \tag{A.1}$$

Note that

$$\sum_l k_l^{\text{in}} = \sum_l k_l^{\text{out}} \equiv \sum_l k_l. \tag{A.2}$$

Here me made the definition that the degree k without superscription refers to in- *or* out-degrees. Without correlations between in- and out-degrees one can perform a coarse-graining of the network as in Sec. 2.1. Thereby one requires the conservation of the individual degrees. As a result, the elements of the approximated adjacency matrix read

$$\tilde{A}_{ij} = k_i^{\text{out}} \frac{k_j^{\text{in}}}{\sum_{l=1}^{N} k_l}. \tag{A.3}$$

Inserting this into the Kuramoto model, Eq. (2.1), one obtains a mean-field description:

$$\dot{\phi}_i(t) = \omega_i + r(t)K\frac{k_i^{\text{out}}}{N}\sin\left[\Theta(t) - \phi_i(t)\right] + \xi_i(t), \tag{A.4}$$

where the mean-field amplitude $r(t)$ and phase $\Theta(t)$ are given by

$$r(t)\mathrm{e}^{i\Theta(t)} := \frac{\sum_l k_l^{\text{in}} \mathrm{e}^{i\phi_l(t)}}{\sum_l k_l}. \tag{A.5}$$

In the following, we consider the thermodynamic limit $N \to \infty$, where the system is conveniently described by the density $\rho(\phi, t|\omega, k^{\text{in}}, k^{\text{out}})$, which is normalized according to $\int_0^{2\pi} \rho(\phi, t|\omega, k^{\text{in}}, k^{\text{out}})d\phi = 1 \ \forall \ \omega, k^{\text{in}}, k^{\text{out}}, t$.

For given in- and out-degree, k^{in} and k^{out}, and natural frequency ω, $\rho(\phi, t|\omega, k^{\text{in}}, k^{\text{out}})$ gives the fraction of oscillators having a phase between ϕ and $\phi + d\phi$ at time t (indices can be neglected, since oscillators with the same individual quantities are assumed to be statistically identical).

The completely asynchronous state is given by $\rho(\phi, t|\omega, k^{\text{in}}, k^{\text{out}}) = 1/(2\pi)\ \forall\, \omega, k^{\text{in}}, k^{\text{out}}, t$ and we aim at calculating the critical coupling strength K_c, where it loses its linear stability, which marks the onset of synchronization.

Analogously to Sec. 2.3, via a linear stability analysis of the completely asynchronous state and invoking a self-consistency condition, the critical coupling strength can be calculated:

$$K_c = 2N\langle k'\rangle \left[\int_{-\infty}^{+\infty} d\omega \sum_{k^{\text{in}},k^{\text{out}}} \frac{Dk^{\text{in}}k^{\text{out}}}{D^2+\omega^2} P\left(\omega, k^{\text{in}}, k^{\text{out}}\right)\right]^{-1} . \tag{A.6}$$

This result generalizes our previously derived expression, Eq. (2.41). The double sum covers all possible in- and out-degrees, which could be further approximated by two integrals. The joint probability density $P\left(\omega, k^{\text{in}}, k^{\text{out}}\right)$ takes into account the possibility of dependencies among the three individual quantities. In the derivation of (A.6) one assumes that, with regard to the ω-dependency, $P\left(\omega, k^{\text{in}}, k^{\text{out}}\right)$ has a single maximum at frequency $\omega = 0$ (this choice is always possible due to the rotational symmetry) and is symmetric with respect to it. Here we neglect degree-frequency correlations. Therefore, we can take $P\left(\omega, k^{\text{in}}, k^{\text{out}}\right) = g(\omega)P\left(k^{\text{in}}, k^{\text{out}}\right)$, where $g(\omega)$ is some given frequency distribution. As a consequence, one achieves a separation into topology and diversity functionals, see Sec. 2.3:

$$K_c = 2\left[\int_{-\infty}^{+\infty} d\omega \frac{D}{D^2+\omega^2} g\left(\omega\right)\right]^{-1} \frac{N\langle k'\rangle}{\langle k^{\text{in}}k^{\text{out}}\rangle} . \tag{A.7}$$

This formula can be seen as a consistent generalization of results presented in [198] towards additive noise. With the standard correlation coefficient "Pearson's R" (we use a capital letter not to confuse it with the Kuramoto order parameter), one can rewrite the average over in- and out-degrees,

$$\langle k^{\text{in}}k^{\text{out}}\rangle = \langle k'\rangle^2 + R\sqrt{\left(\left\langle (k^{\text{in}})^2\right\rangle - \langle k'\rangle^2\right)\left(\left\langle (k^{\text{out}})^2\right\rangle - \langle k'\rangle^2\right)}. \tag{A.8}$$

Pearson's R measures the correlation between in- and out-degrees and lies between -1 and $+1$. The values ± 1 correspond to perfect linear dependencies with arbitrary positive or negative slopes, respectively. Plugging (A.8) into (A.7), we see that positive correlations decrease the critical coupling strength and hence make it easier for the network to synchronize, while the opposite is true for negative correlations between in- and out-degrees. We mention two real-world examples for positive correlations: email networks with $R = 0.53$ [199] and the C. elegans neuronal network with $R = 0.52$ [72]. It will be interesting to further delve into the effects of such correlations, specifically in dependence of the natural frequencies.

B. Beyond uncorrelated networks

Differently than in Sec. A, here we come back to undirected networks and discuss the extension to another important topic. Specifically, it has to be acknowledged that real-world systems are in general structured. It is possible to go beyond the mean-field approximation that is based on coarse-graining a random network of oscilllators with uncorrelated individual quantities (see Secs. 2.1-2.3). To this end, one performs a linear stability analysis of the incoherent state on the level of the phase variables themselves [cf. Ref. [57]]. First it is assumed that in the incoherent state the phases evolve approximately according to their natural frequency ω_i perturbed by the additive Gaussian white noise $\xi_i(t)$:

$$\phi_i^{\text{incoh}}(t) \approx \phi_i^0 + \omega_i t + \xi_i(t), \; i = 1, \dots, N, \tag{B.1}$$

where ϕ_i^0 is the initial phase of oscillator i, and the noise obeys $\langle \xi_i(t) \rangle = 0$ and $\langle \xi_i(t)\xi_j(t') \rangle = 2D\delta_{ij}\delta(t-t')$. Next, infinitesimal perturbations $\delta_i(t)$ are added in Eq. (B.1) which are assumed to grow in time exponentially $\sim \exp(\lambda t)$:

$$\phi_i(t) = \phi_i^{\text{incoh}}(t) + \delta_i(t). \tag{B.2}$$

With this, a linear stability approach can be used. Neglecting terms that are small due to incoherence and due to sufficiently large degrees k_i [57], one can derive the following eigenvalue equation with a set of N constants B_i, $i = 1, \dots, N$ that need to be determined self-consistently:

$$B_i = \frac{K}{2N} \sum_{j=1}^{N} \frac{A_{ij} B_j}{\lambda + D + i\omega_j}. \tag{B.3}$$

In principle, this equation is more general than the similar one derived in Sec. 2.3 through a different approach, see Eq. (2.38), because (B.3) retains the full adjacency matrix. Assuming solutions for B_i independent of the natural frequencies ω_i and assuming no correlations between the network structure and the frequencies, $(\lambda + D + i\omega_j)^{-1}$ can be replaced by its expected value such that the eigenvalue [growth rate of small perturbations, cf. Eq. (B.2)] is given in the limit $N \to \infty$ by

$$1 = \frac{K\lambda_A^{\max}}{2N} \int_{-\infty}^{\infty} d\omega' \frac{g(\omega')}{\lambda + D + i\omega'}, \tag{B.4}$$

where $\lambda_A^{\max}$ is the largest eigenvalue of the adjacency matrix. If we consider, as in Sec. 2.3, a unimodal and symmetric frequency distribution $g(\omega)$, then there is only one solution for λ and it has to be real-valued. Hence, we obtain the critical coupling strength for the onset of synchronization as

$$K_c = \frac{2N}{\lambda_A^{\max}} \left[\int_{-\infty}^{+\infty} d\omega' \frac{Dg(\omega')}{D^2 + \omega'^2} \right]^{-1} \tag{B.5}$$

This result resembles our previous result, Eqs. (2.42)-(2.44), except that instead of the ratio of the first two moments of the degree distribution one has a rescaling via $\lambda_A^{\max}$, the largest eigenvalue of the adjacency matrix. With Eq. (B.5), the effect of degree correlations on the critical coupling can be studied in principle. The drawback is that it is not possible to make further analytical progress. We aim for a mean-field description which is generated by a network coarse-graining that goes beyond Sec. 2.1. In the next sections we present the first steps and show the potential of this approach to obtain new theoretical insights.

B.1. Two-point correlated random networks

Departing from the assumption of uncorrelated individual quantities, the most relevant correlations are the ones that span two nodes, so-called two-point correlations. In order to establish a general framework for the study of collective dynamics taking place on arbitrarily two-point correlated random networks, we introduce the following ansatz

$$A_{ij} = \frac{k_i k_j C_{ij}}{\sum_{l=1}^{N} k_l}. \tag{B.6}$$

The constants C_{ij} may encode various two-point correlations (degrees, frequencies, initial conditions etc.). In the neutral case $C_{ij} = 1 \, \forall \, i, j$, the elements of the adjacency matrix are approximated by the probability for the connection in an uncorrelated random network, cf. Sec. 2.1. From the general fact $\sum_{i,j} A_{ij} = \sum_l k_l$ it follows that a constraint on the C_{ij} must be fulfilled, namely $\sum_{i,j} k_i k_j C_{ij} / \left(\sum_l k_l\right)^2 \stackrel{!}{=} 1$.

Here we use Eq. (B.6) to replace the adjacency matrix in the stochastic Kuramoto model. Furthermore, by defining the following local mean-field amplitudes and phases,

$$r_i(t) \mathrm{e}^{i\Theta_i(t)} := \frac{\sum_j C_{ij} k_j \mathrm{e}^{i\phi_j(t)}}{\sum_j k_j}, \tag{B.7}$$

the original system (2.1) can be brought into a mean-field model:

$$\dot{\phi}_i(t) = \omega_i + \frac{K}{N} k_i r_i(t) \sin\left[\Theta_i(t) - \phi_i(t)\right] + \xi_i(t). \tag{B.8}$$

Let us focus on two-point correlations with respect to natural frequencies and degrees. Given an oscillator network with a set of natural frequencies ω_i, $i = 1, \ldots, N$ and a set of degrees k_i, $i = 1, \ldots, N$, the underlying question is, how the critical coupling strength for the onset of synchronization is affected by the way the connectivity is chosen between the oscillators. We note that one could proceed by inserting Eq. (B.6) into Eq. (B.3), but as we explained above, we are interested here in a theory that enables more explicit insights. Inspired by works of epidemic spreading on correlated networks [200], we consider the following correlation matrix

$$C_{ij} = \frac{P_e(\omega_j, k_j | \omega_i, k_i)}{P_e(\omega_j, k_j)}. \tag{B.9}$$

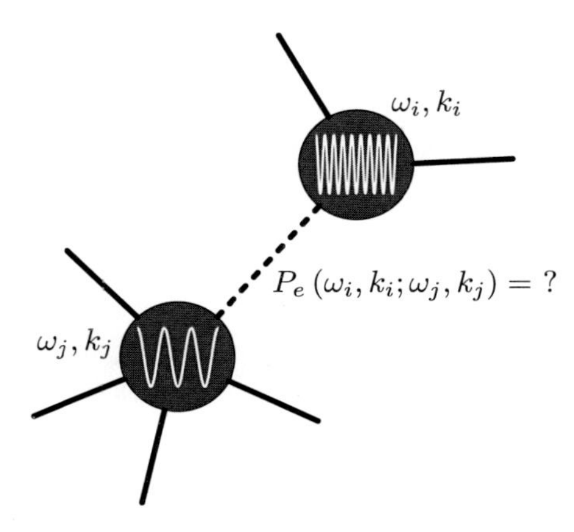

Figure B.1.: Sketch of a random network with potential correlations in the natural frequencies and degrees between two arbitrary connected nodes. The probability to find oscillators with frequencies ω_i, ω_j and degrees k_i, k_j at the ends of a randomly selected edge, is given by $P_e(\omega_i, k_i; \omega_j, k_j)$.

Here we use edge-based probability distributions denoted by the index e. They have to be distinguished from node-based probability distributions that so far were used throughout our work. Indeed, picking randomly a node from the network and asking for the probability that this node has a certain degree and frequency, is different from picking randomly an edge and asking for the probability that there is a node at the ends of that edge with a certain degree and frequency. Specifically, the edge- and node-based frequency-degree probability distributions are related via

$$P_e(\omega_i, k_i) = \frac{k_i}{\langle k \rangle} P(\omega_i, k_i). \tag{B.10}$$

The distribution $P_e(\omega_j, k_j|\omega_i, k_i)$ gives the probability that for given node with frequency ω_i and degree k_i at the end of a randomly selected edge, at the other end there is a node with frequency ω_j and degree k_j. Using Eq. (B.10) and the connection between joint and conditional probability distributions, $P_e(\omega_j, k_j|\omega_i, k_i) = P_e(\omega_j, k_j; \omega_i, k_i)/P_e(\omega_i, k_i)$, the aforementioned constraint translates into

$$\frac{1}{N^2} \sum_{i,j} \frac{P_e(\omega_i, k_i; \omega_j, k_j)}{P(\omega_i, k_i) P(\omega_j, k_j)} \stackrel{!}{=} 1. \tag{B.11}$$

For a visualization of the random networks that we have in mind here, see Fig. B.1. We use the equations above and proceed as in Secs. 2.2 and 2.3. In particular, we assume again that dynamical correlations between any two oscillators can be neglected, cf. Eq. (2.17). Then the evolution equation for the fundamental mode $c_1(t|\omega, k)$ reads [compare with Eq. (2.34)]:

$$\frac{\partial c_1}{\partial t} = \frac{Kk}{2N} \int_{-\infty}^{+\infty} d\omega' \sum_{k'=k_{\min}}^{k_{\max}} c_1(t|\omega', k') P_e(\omega', k'|\omega, k) - (D + i\omega) c_1 \ . \tag{B.12}$$

We are now in the position to consider specific examples.

B.2. Assortativity by degree

As a first example we consider only degree-degree correlations [(dis)assortativity refer to (negative) positive correlations]. We take identical frequencies, $\omega_i = 0 \ \forall i$. Then the evolution equation for the fundamental mode becomes

$$\begin{aligned} \dot{c_1}(t|k) =& -Dc_1(t|k) + \frac{Kk}{2N} \sum_{k'=k_{\min}}^{k_{\max}} c_1(t|k') P_e(k'|k) \\ \equiv& \sum_{k'=k_{\min}}^{k_{\max}} J_{kk'} c_1(t|k') \ . \end{aligned} \tag{B.13}$$

Here we defined the Jacobian matrix with entries $J_{kk'} = -D\delta_{kk'} + \frac{Kk}{2N} P_e(k'|k)$. Let $\lambda_P^{\max}$ be the largest eigenvalue of $kP_e(k'|k)$. Then from Eq. (B.13) it follows the critical coupling strength,

$$K_c = \frac{2DN}{\lambda_P^{\max}}. \tag{B.14}$$

One may argue from this, similarly as in Ref. [200], that assortativity by degree tends to decrease the critical coupling. In other words, connecting preferentially similar nodes (hubs with hubs and non-hubs with non-hubs) makes it easier for the oscillator network to partially synchronize.

B.3. Assortativity by frequency

We proceed with frequency distributions $g(\omega)$ that have compact support, $\omega \in [0, \gamma]$ (note that due to the rotational symmetry this interval can be shifted without changing the collective dynamics). We further propose to discretize the frequencies as $\gamma \equiv N\Delta\omega$ to study frequency-frequency correlations. Let us consider here regular networks where all nodes have the same degree. Then we obtain for the real and imaginary parts of the fundamental mode the following evolution equations:

$$\begin{aligned} \dot{c}_\nu^R =& \frac{K}{2} \sum_{\nu'=0}^{N} P_e(\nu'\Delta\omega|\nu\Delta\omega)\, c_{\nu'}^R - Dc_\nu^R + \nu\Delta\omega c_\nu^I \\ \dot{c}_\nu^I =& \frac{K}{2} \sum_{\nu'=0}^{N} P_e(\nu'\Delta\omega|\nu\Delta\omega)\, c_{\nu'}^I - Dc_\nu^I - \nu\Delta\omega c_\nu^R. \end{aligned} \tag{B.15}$$

The Jacobian for this problem can be written in form of a block matrix:

$$\mathbf{J} = \begin{pmatrix} G & H \\ -H & G \end{pmatrix}, \tag{B.16}$$

with $G_{\nu\nu'} \equiv -D\delta_{\nu\nu'} + (K/2)P_e\,(\nu'\Delta\omega|\nu\Delta\omega)$ and $H_{\nu\nu'} \equiv \nu\Delta\omega\delta_{\nu\nu'}$. Making use of the fact that for the determinant of a block matrix one has $\det\left(\begin{smallmatrix} A & B \\ C & D \end{smallmatrix}\right) = \det\left(AD - BC\right)$, if A and B commute, we finally obtain the characteristic equation for the eigenvalues λ:

$$\det\left[(D+\lambda)^2 I - K(D+\lambda)\Omega + \frac{K^2}{4}\Omega^2 + W\right] = 0, \tag{B.17}$$

where I is the identity matrix and for ease of notation $\Omega_{\nu\nu'} = P_e\,(\nu'\Delta\omega|\nu\Delta\omega)$ and $W_{\nu\nu'} = \nu^2\,(\Delta\omega)^2\,\delta_{\nu\nu'}$. We call Ω the mixing matrix.

For a perfectly assortative mixing $\Omega_{\nu\nu'} = \delta_{\nu\nu'}$, we find from Eq. (B.17)

$$\lambda_n = \frac{K}{2} - D \pm in\Delta\omega, \tag{B.18}$$

$n = 1, 2, \ldots, N$. Hence, all eigenvalues have the same real part, and setting it to zero gives the critical coupling strength $K_c = 2D$, irrespective of the spread in the natural frequencies. This finding differs strongly from what is known for uncorrelated networks, where the critical coupling strength is increased by frequency mismatches. The validity of the theory presented here, along with further interesting examples using Eq. (B.17), shall be considered elsewhere. We expect that our approach here is useful to better understand the collective dynamics in correlated neuronal networks [201].

Bibliography

[1] S. Strogatz. *Sync: The Emerging Science of Spontaneous Order.* Hyperion, New York, 2003.

[2] A. Pikovsky, M. Rosenblum, and J. Kurths. *Synchronization: A universal concept in nonlinear sciences.* Cambridge Univ. Press, U. K., 2003.

[3] A. K. Engel, P. Fries, P. König, M. Brecht, and W. Singer. Temporal binding, binocular rivalry, and consciousness. *Consciousness and cognition*, 8(2):128–151, 1999.

[4] J. F. Hipp, A. K. Engel, and M. Siegel. Oscillatory synchronization in large-scale cortical networks predicts perception. *Neuron*, 69(2):387–396, 2011.

[5] T. I. Netoff, R. Clewley, S. Arno, T. Keck, and J. A. White. Epilepsy in small-world networks. *J. Neurosci.*, 24(37):8075–8083, 2004.

[6] P. J. Uhlhaas, G. Pipa, B. Lima, L. Melloni, S. Neuenschwander, D. Nikolić, and W. Singer. Neural Synchrony in Cortical Networks: History, Concept and Current Status. *Front. Integr. Neurosci.*, 3(17):1–19, 2009.

[7] O. Sporns, D. R. Chialvo, M. Kaiser, and C. C. Hilgetag. Organization, development and function of complex brain networks. *Trends Cogn. Sci.*, 8(9):418–425, 2004.

[8] G. Buzsáki. *Rhythms of the Brain.* Oxford University Press, 2006.

[9] O. Sporns. *Networks Of The Brain.* MIT Press, 2010.

[10] C. R. Gallistel and A. P. King. *Memory and the Computational Brain.* Wiley-Blackwell, 2010.

[11] O. Sporns, G. Tononi, and R. Kötter. The human connectome: A structural description of the human brain. *PLoS Computational Biology*, 1(4):245–251, 2005.

[12] H. Haken. *Principles of brain functioning: a synergetic approach to brain activity, behavior and cognition*, volume 67. Springer Science & Business Media, New York, 2013.

[13] A. T. Winfree. Biological rhythms and the behavior of populations of coupled oscillators. *J. Theor. Biol.*, 16(1):15–42, 1967.

[14] Y. Kuramoto. Self-entrainment of a population of coupled non-linear oscillators. In H. Araki, editor, *Lect. Notes Phys.*, volume 39, pages 420–422. Springer, New York, 1975.

[15] Y. Kuramoto. *Chemical Oscillations, Waves, and Turbulence.* Springer-Verlag, Berlin, 1984.

[16] S. H. Strogatz. From Kuramoto to Crawford: exploring the onset of synchronization in populations of coupled oscillators. *Physica D*, 143:1–20, 2000.

[17] J. A. Acebrón, L. L. Bonilla, C. J. Pérez-Vicente, F. Ritort, and R. Spigler. The Kuramoto model: A simple paradigm for synchronization phenomena. *Rev. Mod. Phys.*, 77:137–185, 2005.

[18] D. Cumin and C. P. Unsworth. Generalising the Kuramoto model for the study of neuronal synchronisation in the brain. *Physica D*, 226(2):181, 2007.

[19] M. Breakspear, S. Heitmann, and A. Daffertshofer. Generative Models of Cortical Oscillations: Neurobiological Implications of the Kuramoto Model. *Front. Hum. Neurosci.*, 4:190, 2010.

[20] F. C. Hoppensteadt and E. M. Izhikevich. *Weakly connected neural networks*, volume 126. Springer Science & Business Media, New York, 2012.

[21] R. M. Smeal, G. B. Ermentrout, and J. A. White. Phase-response curves and synchronized neural networks. *Phil. Trans. R. Soc. B*, 365(1551):2407–2422, 2010.

[22] B. Lindner, J. García-Ojalvo, A. Neiman, and L. Schimansky-Geier. Effects of noise in excitable systems. *Phys. Rep.*, 392:321–424, 2004.

[23] R. S. Fisher, W. van Emde Boas, W. Blume, C. Elger, P. Genton, P. Lee, and J. Engel. Epileptic seizures and epilepsy: definitions proposed by the International League Against Epilepsy (ILAE) and the International Bureau for Epilepsy (IBE). *Epilepsia*, 46(4):470–472, 2005.

[24] C. Hammond, H. Bergman, and P. Brown. Pathological synchronization in Parkinson's disease: networks, models and treatments. *Trends Neurosci.*, 30(7):357–364, 2007.

[25] K. M. Spencer, P. G. Nestor, M. A. Niznikiewicz, D. F. Salisbury, M. E. Shenton, and R. W. McCarley. Abnormal neural synchrony in schizophrenia. *J. Neurosci.*, 23(19):7407–7411, 2003.

[26] P. J. Uhlhaas and W. Singer. Abnormal neural oscillations and synchrony in schizophrenia. *Nature Rev. Neurosci.*, 11(2):100–113, 2010.

[27] C. J. Stam, B. F. Jones, G. Nolte, M. Breakspear, and Ph. Scheltens. Small-world networks and functional connectivity in Alzheimer's disease. *Cereb. Cortex*, 17(1):92–99, 2007.

[28] P. J. Uhlhaas and W. Singer. Neural synchrony in brain disorders: relevance for cognitive dysfunctions and pathophysiology. *Neuron*, 52(1):155–168, 2006.

[29] C. W. Wu. *Synchronization in Complex Networks of Nonlinear Dynamical Systems*. World Scientific, 2007.

[30] A. Arenas, A. Díaz-Guilera, J. Kurths, Y. Moreno, and C. Zhou. Synchronization in complex networks. *Phys. Rep.*, 469(3):93–153, 2008.

[31] J. N. Teramae, H. Nakao, and G. B. Ermentrout. Stochastic phase reduction for a general class of noisy limit cycle oscillators. *Phys. Rev. Lett.*, 102(19):194102, 2009.

[32] D. S. Goldobin, J. N. Teramae, H. Nakao, and G. B. Ermentrout. Dynamics of limit-cycle oscillators subject to general noise. *Phys. Rev. Lett.*, 105(15):154101, 2010.

[33] J. T. C. Schwabedal and A. Pikovsky. Effective phase dynamics of noise-induced oscillations in excitable systems. *Phys. Rev. E*, 81(4):046218, 2010.

[34] J. T. C. Schwabedal and A. Pikovsky. Effective phase description of noise-perturbed and noise-induced oscillations. *Eur. Phys. J. Special Topics*, 187(1):63–76, 2010.

[35] Y. Kawamura, H. Nakao, and Y. Kuramoto. Collective phase description of globally coupled excitable elements. *Phys. Rev. E*, 84(4):046211, 2011.

[36] J. T. C. Schwabedal and A. Pikovsky. Phase description of stochastic oscillations. *Phys. Rev. Lett.*, 110(20):204102, 2013.

[37] P. J. Thomas and B. Lindner. Asymptotic Phase for Stochastic Oscillators. *Phys. Rev. Lett.*, 113(25):254101, 2014.

[38] A. Pikovsky. Comment on "Asymptotic Phase for Stochastic Oscillators". *Phys. Rev. Lett.*, 115(6):069401, 2015.

[39] P. J. Thomas and B. Lindner. Thomas and Lindner Reply. *Phys. Rev. Lett.*, 115(6):069402, 2015.

[40] R. Pastor-Satorras and A. Vespignani. Epidemic dynamics and endemic states in complex networks. *Phys. Rev. E*, 63(6):066117, 2001.

[41] T. Ichinomiya. Frequency synchronization in a random oscillator network. *Phys. Rev. E*, 70(2):026116, 2004.

[42] D.-S. Lee. Synchronization transition in scale-free networks: Clusters of synchrony. *Phys. Rev. E*, 72(2):026208, 2005.

[43] V. Sood and S. Redner. Voter model on heterogeneous graphs. *Phys. Rev. Lett.*, 94(17):178701, 2005.

[44] A. Barrat, M. Barthélemy, and A. Vespignani. *Dynamical Processes on Complex Networks.* Cambridge Univ. Press, U. K., 2008.

[45] F. Chung and L. Lu. Connected components in random graphs with given expected degree sequences. *Annals of combinatorics*, 6(2):125–145, 2002.

[46] H. Risken. *The Fokker-Planck Equation.* Springer, 1996.

[47] L. L. Bonilla, C. J. Pérez Vicente, and J. M. Rubi. Glassy synchronization in a population of coupled oscillators. *J. Stat. Phys.*, 70(3-4):921–937, 1993.

[48] J. D. Crawford and K. T. R. Davies. Synchronization of globally coupled phase oscillators: singularities and scaling for general couplings. *Physica D*, 125(1-2):1–46, 1999.

[49] E. J. Hildebrand, M. A. Buice, and C. C. Chow. Kinetic Theory of Coupled Oscillators. *Phys. Rev. Lett.*, 98:054101, 2007.

[50] O. Faugeras, J. Touboul, and B. Cessac. A Constructive Mean-Field Analysis of Multi-Population Neural Networks with Random Synaptic Weights and Stochastic Inputs. *Front. Comput. Neurosci.*, 3:1, 2009.

[51] S. Mischler and C. Mouhot. Kac's program in kinetic theory. *Invent. math.*, 193(1):1–147, 2013.

[52] M. Kac. Foundations of kinetic theory. In *Proc. Third Berkeley Symp. on Math. Statist. and Prob.*, volume 3, pages 171–197. Univ. of Calif. Press, 1956.

[53] T. D. Frank. *Nonlinear Fokker-Planck equations: fundamentals and applications.* Springer, 2005.

[54] S. H. Strogatz and R. E. Mirollo. Stability of incoherence in a population of coupled oscillators. *J. Stat. Phys.*, 63(3-4):613, 1991.

[55] E. Ott. *Chaos in Dynamical Systems.* Cambridge Univ. Press, U. K., 2002.

[56] R. E. Mirollo and S. H. Strogatz. Amplitude death in an array of limit-cycle oscillators. *J. Stat. Phys.*, 60(1-2):245–261, 1990.

[57] J. G. Restrepo, E. Ott, and B. R. Hunt. Onset of synchronization in large networks of coupled oscillators. *Phys. Rev. E*, 71(3):036151, 2005.

[58] B. R. Preiss. *Data Structures and Algorithms with Object-Oriented Design Patterns in C++.* Wiley & Sons, 1998.

[59] R. Albert and A.-L. Barabási. Statistical mechanics of complex networks. *Rev. Mod. Phys.*, 74(1):47–97, 2002.

[60] M. E. J. Newman. The structure and function of complex networks. *SIAM Rev.*, 45(2):167–256, 2003.

[61] S. Boccaletti, V. Latora, Y. Moreno, M. Chavez, and D.-U. Hwang. Complex networks: Structure and dynamics. *Phys. Rep.*, 424(4):175–308, 2006.

[62] D. J. Watts and S. H. Strogatz. Collective dynamics of 'small-world' networks. *Nature*, 393:440–442, 1998.

[63] B. Sonnenschein and L. Schimansky-Geier. Onset of synchronization in complex networks of noisy oscillators. *Phys. Rev. E*, 85:051116, 2012.

[64] H. F. Song and X.-J. Wang. Simple, distance-dependent formulation of the Watts-Strogatz model for directed and undirected small-world networks. *Phys. Rev. E*, 90(6):062801, 2014.

[65] M. E. J. Newman and D. J. Watts. Scaling and percolation in the small-world network model. *Phys. Rev. E*, 60(6):7332–7342, 1999.

[66] R. Mannella. A Gentle Introduction to the Integration of Stochastic Differential Equations. In J. A. Freund and T. Pöschel, editors, *Stochastic Processes in Physics, Chemistry, and Biology*, volume 557, pages 353–364. Springer-Verlag, Berlin Heidelberg, 2000.

[67] R. Toral and P. Colet. *Stochastic Numerical Methods.* Wiley-VCH, 2014.

[68] H. Hong, M. Y. Choi, and B. J. Kim. Synchronization on small-world networks. *Phys. Rev. E*, 65(2):026139, 2002.

[69] S. W. Son and H. Hong. Thermal fluctuation effects on finite-size scaling of synchronization. *Phys. Rev. E*, 81(6):061125, 2010.

[70] C. Choi, M. Ha, and B. Kahng. Extended finite-size scaling of synchronized coupled oscillators. *Phys. Review E*, 88(3):032126, 2013.

[71] M. J. Lee, S. Do Yi, and B. J. Kim. Finite-Time and Finite-Size Scaling of the Kuramoto Oscillators. *Phys. Rev. Lett.*, 112(7):074102, 2014.

[72] L. R. Varshney, B. L. Chen, E. Paniagua, D. H. Hall, and D. B. Chklovskii. Structural properties of the Caenorhabditis elegans neuronal network. *PLoS Comput. Biol.*, 7(2):e1001066, 2011.

[73] A. Balanov, N. Janson, D. Postnov, and O. Sosnovtseva. *Synchronization: From Simple to Complex.* Springer-Verlag, Berlin, 2010.

[74] M. Brede. Synchrony-optimized networks of non-identical Kuramoto oscillators. *Phys. Lett. A*, 372(15):2618, 2008.

[75] J. Fan and D. J. Hill. Enhancement of Synchronizability of the Kuramoto Model with Assortative Degree-Frequency Mixing. In J. Zhou, editor, *Complex Sciences*, volume 5, pages 1967–1972. Springer-Verlag, Berlin Heidelberg, 2009.

[76] J. Gómez-Gardenes, S. Gómez, A. Arenas, and Y. Moreno. Explosive Synchronization Transitions in Scale-free Networks. *Phys. Rev. Lett.*, 106:128701, 2011.

[77] T. P. Vogels, H. Sprekeler, F. Zenke, C. Clopath, and W. Gerstner. Inhibitory Plasticity Balances Excitation and Inhibition in Sensory Pathways and Memory Networks. *Science*, 334(6062):1569, 2011.

[78] B. Sonnenschein, F. Sagués, and L. Schimansky-Geier. Networks of noisy oscillators with correlated degree and frequency dispersion. *Eur. Phys. J. B*, 86:12, 2013.

[79] T. Pereira. Hub synchronization in scale-free networks. *Phys. Rev. E*, 82(3):036201, 2010.

[80] H. Sakaguchi. Cooperative Phenomena in Coupled Oscillator Systems under External Fields. *Prog. Theor. Phys.*, 79(1):39, 1988.

[81] P. S. Skardal, J. Sun, D. Taylor, and J. G. Restrepo. Effects of degree-frequency correlations on network synchronization: Universality and full phase-locking. *Eur. Phys. Lett.*, 101(2):20001, 2013.

[82] Y. Kuramoto and I. Nishikawa. Statistical macrodynamics of large dynamical systems. Case of a phase transition in oscillator communities. *J. Stat. Phys.*, 49(3-4):569, 1987.

[83] S. H. Strogatz, R. E. Mirollo, and P. C. Matthews. Coupled nonlinear oscillators below the synchronization threshold: relaxation by generalized landau damping. *Phys. Rev. Lett.*, 68(18):2730, 1992.

[84] S. Watanabe and S. H. Strogatz. Integrability of a globally coupled oscillator array. *Phys. Rev. Lett.*, 70(16):2391, 1993.

[85] S. Watanabe and S. H. Strogatz. Constants of motion for superconducting josephson arrays. *Physica D*, 74(3):197–253, 1994.

[86] A. Pikovsky and M. Rosenblum. Partially integrable dynamics of hierarchical populations of coupled oscillators. *Phys. Rev. Lett.*, 101(26):264103, 2008.

[87] A. Pikovsky and M. Rosenblum. Dynamics of heterogeneous oscillator ensembles in terms of collective variables. *Physica D*, 240(9):872–881, 2011.

[88] E. A. Martens, E. Barreto, S. H. Strogatz, E. Ott, P. So, and T. M. Antonsen. Exact results for the Kuramoto model with a bimodal frequency distribution. *Phys. Rev. E*, 79(2):026204, 2009.

[89] S. A. Marvel, R. E. Mirollo, and S. H. Strogatz. Identical phase oscillators with global sinusoidal coupling evolve by Möbius group action. *Chaos*, 19(4):043104, 2009.

[90] E. Ott and T. M. Antonsen. Low dimensional behavior of large systems of globally coupled oscillators. *Chaos*, 18(3):037113, 2008.

[91] E. Ott and T. M. Antonsen. Long time evolution of phase oscillator systems. *Chaos*, 19(2):023117, 2009.

[92] G. Giacomin, K. Pakdaman, and X. Pellegrin. Global attractor and asymptotic dynamics in the Kuramoto model for coupled noisy phase oscillators. *Nonlinearity*, 25(5):1247, 2012.

[93] R. E. Mirollo. The asymptotic behavior of the order parameter for the infinite-N Kuramoto model. *Chaos*, 22(4):043118, 2012.

[94] O. E. Omel'chenko and M. Wolfrum. Bifurcations in the Sakaguchi-Kuramoto model. *Physica D*, 263:74–85, 2013.

[95] K. H. Nagai and H. Kori. Noise-induced synchronization of a large population of globally coupled nonidentical oscillators. *Phys. Rev. E*, 81(6):065202(R), 2010.

[96] W. Braun, A. Pikovsky, M. A. Matias, and P. Colet. Global dynamics of oscillator populations under common noise. *Europhys. Lett.*, 99(2):20006, 2012.

[97] Y. M. Lai and M. A. Porter. Noise-induced synchronization, desynchronization, and clustering in globally coupled nonidentical oscillators. *Phys. Rev. E*, 88(1):012905, 2013.

[98] C. Kurrer and K. Schulten. Noise-induced synchronous neuronal oscillations. *Phys. Rev. E*, 51(6):6213, 1995.

[99] M. A. Zaks, A. B. Neiman, S. Feistel, and L. Schimansky-Geier. Noise-controlled oscillations and their bifurcations in coupled phase oscillators. *Phys. Rev. E*, 68(6):066206, 2003.

[100] V. S. Anishchenko, V. Astakhov, A. Neiman, T. Vadivasova, and L. Schimansky-Geier. *Nonlinear Dynamics of Chaotic and Stochastic Systems.* Springer-Verlag, Berlin, 2007.

[101] H. Sakaguchi, S. Shinomoto, and Y. Kuramoto. Phase transitions and their bifurcation analysis in a large population of active rotators with mean-field coupling. *Prog. Theor. Phys.*, 79(3):600, 1988.

[102] Kanti V Mardia and Peter E Jupp. *Directional statistics*, volume 494. John Wiley & Sons, 2009.

[103] J. D. Crawford. Amplitude expansions for instabilities in populations of globally-coupled oscillators. *J. Stat. Phys.*, 74(5-6):1047, 1994.

[104] A. Arenas and C. J. Pérez Vicente. Exact long-time behavior of a network of phase oscillators under random fields. *Phys. Rev. E*, 50(2):949, 1994.

[105] J. D. Crawford. Scaling and Singularities in the Entrainment of Globally Coupled Oscillators. *Phys. Rev. Lett.*, 74(21):4341, 1995.

[106] A. Pikovsky and S. Ruffo. Finite-size effects in a population of interacting oscillators. *Phys. Rev. E*, 59(2):1633, 1999.

[107] N. J. Balmforth and R. Sassi. A shocking display of synchrony. *Physica D*, 143(1):21–55, 2000.

[108] C. J. Tessone, A. Scirè, R. Toral, and P. Colet. Theory of collective firing induced by noise or diversity in excitable media. *Phys. Rev. E*, 75(1):016203, 2007.

[109] L. Bertini, G. Giacomin, and K. Pakdaman. Dynamical Aspects of Mean Field Plane Rotators and the Kuramoto Model. *J. Stat. Phys.*, 138(1-3):270–290, 2010.

[110] B. Sonnenschein and L. Schimansky-Geier. Approximate solution to the stochastic Kuramoto model. *Phys. Rev. E*, 88(5):052111, 2013.

[111] D. Hennig, C. Mulhern, L. Schimansky-Geier, G. P. Tsironis, and P. Hänggi. Cooperative surmounting of bottlenecks. *Phys. Rep.*, 586:1–51, 2015.

[112] S. Shinomoto and Y. Kuramoto. Phase transitions in active rotator systems. *Prog. Theor. Phys.*, 75(5):1105, 1986.

[113] M. Kostur, J. Luczka, and L. Schimansky-Geier. Nonequilibrium coupled Brownian phase oscillators. *Phys. Rev. E*, 65(5):051115, 2002.

[114] L. M. Childs and S. H. Strogatz. Stability diagram for the forced Kuramoto model. *Chaos*, 18:043128, 2008.

[115] L. F. Lafuerza, P. Colet, and R. Toral. Nonuniversal Results Induced by Diversity Distribution in Coupled Excitable Systems. *Phys. Rev. Lett.*, 105(8):084101, 2010.

[116] C. J. Tessone, D. H. Zanette, and R. Toral. Global firing induced by network disorder in ensembles of active rotators. *Eur. Phys. J. B*, 62:319, 2008.

[117] Robert Adler. A study of locking phenomena in oscillators. *Proc. IRE*, 34(6):351–357, 1946.

[118] A. V. Straube and P. Tierno. Synchronous vs. asynchronous transport of a paramagnetic particle in a modulated ratchet potential. *Europhys. Lett.*, 103(2):28001, 2013.

[119] J. A. Kromer, B. Lindner, and L. Schimansky-Geier. Event-triggered feedback in noise-driven phase oscillators. *Phys. Rev. E*, 89(3):032138, 2014.

[120] J. A. Kromer, R. D. Pinto, B. Lindner, and L. Schimansky-Geier. Noise-controlled bistability in an excitable system with positive feedback. *Europhys. Lett.*, 108(2):20007, 2014.

[121] T. Pereira, D. Eroglu, G. B. Bagci, U. Tirnakli, and H. J. Jensen. Connectivity-driven coherence in complex networks. *Phys. Rev. Lett.*, 110(23):234103, 2013.

[122] B. Sonnenschein, M. A. Zaks, A. B. Neiman, and L. Schimansky-Geier. Excitable elements controlled by noise and network structure. *Eur. Phys. J. Special Topics*, 222(10):2517–2529, 2013.

[123] A. Dhooge, W. Govaerts, and Y. A. Kuznetsov. MATCONT: A MATLAB package for numerical bifurcation analysis of ODEs. *ACM Transactions on Mathematical Software*, 29(2):141–164, 2003.

[124] M. E. J. Newman. Assortative mixing in networks. *Phys. Rev. Lett.*, 89(20):208701, 2002.

[125] F. Viger and M. Latapy. Efficient and Simple Generation of Random Simple Connected Graphs with Prescribed Degree Sequence. In L. Wang, editor, *Comput. Combin.*, volume 3595, pages 440–449. Springer Berlin Heidelberg, 2005.

[126] G. Csárdi and T. Nepusz. The igraph software package for complex network research. *InterJournal Complex Systems*, 1695, 2006.

[127] L. Glass. Synchronization and rhythmic processes in physiology. *Nature*, 410(6825):277–284, 2001.

[128] D. E. Postnov, D. D. Postnov, and L. Schimansky-Geier. Self-terminating wave patterns and self-organized pacemakers in a phenomenological model of spreading depression. *Brain research*, 1434:200–211, 2012.

[129] T. B. Luke, E. Barreto, and P. So. Complete classification of the macroscopic behavior of a heterogeneous network of theta neurons. *Neural Comput.*, 25(12):3207–3234, 2013.

[130] S. Luccioli, E. Ben-Jacob, A. Barzilai, P. Bonifazi, and A. Torcini. Clique of Functional Hubs Orchestrates Population Bursts in Developmentally Regulated Neural Networks. *PLoS Comput. Biol.*, 10(9):e1003823, 2014.

[131] D. DiFrancesco. Pacemaker mechanisms in cardiac tissue. *Annu. Rev. Physiol.*, 55(1):455–472, 1993.

[132] T. K. Shajahan, B. Borek, A. Shrier, and L. Glass. Spiralâ€“pacemaker interactions in a mathematical model of excitable medium. *New J. Phys.*, 15(2):023028, 2013.

[133] A. Karma. Physics of cardiac arrhythmogenesis. *Annu. Rev. Condens. Matter Phys.*, 4(1):313–337, 2013.

[134] Z. Qu, G. Hu, A. Garfinkel, and J. N. Weiss. Nonlinear and Stochastic Dynamics in the Heart. *Phys. Rep.*, 543(2):61–162, 2014.

[135] T. K. DM. Peron and F. A. Rodrigues. Determination of the critical coupling of explosive synchronization transitions in scale-free networks by mean-field approximations. *Phys. Rev. E*, 86(5):056108, 2012.

[136] I. Leyva, R. Sevilla-Escoboza, J. M. Buldú, I. Sendiña-Nadal, J. Gómez-Gardeñes, A. Arenas, Y. Moreno, S. Gómez, R. Jaimes-Reátegui, and S. Boccaletti. Explosive first-order transition to synchrony in networked chaotic oscillators. *Phys. Rev. Lett.*, 108:168702, 2012.

[137] B. C. Coutinho, A. V. Goltsev, S. N. Dorogovtsev, and J. F. F. Mendes. Kuramoto model with frequency-degree correlations on complex networks. *Phys. Rev. E*, 87:032106, 2013.

[138] P. Li, K. Zhang, X. Xu, J. Zhang, and M. Small. Reexamination of explosive synchronization in scale-free networks: The effect of disassortativity. *Phys. Rev. E*, 87(4):042803, 2013.

[139] G. Su, Z. Ruan, S. Guan, and Z. Liu. Explosive synchronization on co-evolving networks. *Europhys. Lett.*, 103(4):48004, 2013.

[140] P. Ji, T. K. DM. Peron, P. J. Menck, F. A. Rodrigues, and J. Kurths. Cluster explosive synchronization in complex networks. *Phys. Rev. Lett.*, 110(21):218701, 2013.

[141] L. Zhu, L. Tian, and D. Shi. Criterion for the emergence of explosive synchronization transitions in networks of phase oscillators. *Phys. Rev. E*, 88(4):042921, 2013.

[142] X. Zhang, X. Hu, J. Kurths, and Z. Liu. Explosive synchronization in a general complex network. *Phys. Rev. E*, 88(1):010802(R), 2013.

[143] I. Leyva, A. Navas, I. Sendiña-Nadal, J. A. Almendral, J. M. Buldú, M. Zanin, D. Papo, and S. Boccaletti. Explosive transitions to synchronization in networks of phase oscillators. *Sci. Rep.*, 3:1281, 2013.

[144] L. Zhu, L. Tian, and D. Shi. Explosive transitions to synchronization in weighted static scale-free networks. *Eur. Phys. J. B*, 86:451, 2013.

[145] H. Chen, G. He, F. Huang, C. Shen, and Z. Hou. Explosive synchronization transitions in complex neural networks. *Chaos*, 23(3):033124, 2013.

[146] Y. Zou, T. Pereira, M. Small, Z. Liu, and J. Kurths. Basin of attraction determines hysteresis in explosive synchronization. *Phys. Rev. Lett.*, 112:114102, 2014.

[147] P. S. Skardal and A. Arenas. Disorder induces explosive synchronization. *Phys. Rev. E*, 89(6):062811, 2014.

[148] K. Mikkelsen, A. Imparato, and A. Torcini. Emergence of slow collective oscillations in neural networks with spike-timing dependent plasticity. *Phys. Rev. Lett.*, 110(20):208101, 2013.

[149] J. G. Orlandi, J. Soriano, E. Alvarez-Lacalle, S. Teller, and J. Casademunt. Noise focusing and the emergence of coherent activity in neuronal cultures. *Nat. Phys.*, 9(9):582–590, 2013.

[150] E. Montbrió, J. Kurths, and B. Blasius. Synchronization of two interacting populations of oscillators. *Phys. Rev. E*, 70(5):056125, 2004.

[151] D. M. Abrams, R. Mirollo, S. H. Strogatz, and D. A. Wiley. Solvable model for chimera states of coupled oscillators. *Phys. Rev. Lett.*, 101(8):084103, 2008.

[152] C. R. Laing. Chimera states in heterogeneous networks. *Chaos*, 19:013113, 2009.

[153] H. Hong and S. H. Strogatz. Kuramoto model of coupled oscillators with positive and negative coupling parameters: an example of conformist and contrarian oscillators. *Phys. Rev. Lett.*, 106(5):054102, 2011.

[154] H. Hong and S. H. Strogatz. Conformists and contrarians in a kuramoto model with identical natural frequencies. *Phys. Rev. E*, 84(4):046202, 2011.

[155] D. Pazó and E. Montbrió. Low-dimensional dynamics of populations of pulse-coupled oscillators. *Phys. Rev. X*, 4:011009, 2014.

[156] B. Sonnenschein, T. K. DM. Peron, F. A. Rodrigues, J. Kurths, and L. Schimansky-Geier. Cooperative behavior between oscillatory and excitable units: the peculiar role of positive coupling-frequency correlations. *Eur. Phys. J. B*, 87:182, 2014.

[157] H. Daido and K. Nakanishi. Aging Transition and Universal Scaling in Oscillator Networks. *Phys. Rev. Lett.*, 93(10):104101, 2004.

[158] D. Pazó and E. Montbrió. Universal behavior in populations composed of excitable and self-oscillatory elements. *Phys. Rev. E*, 73:055202(R), 2006.

[159] H. Daido, A. Kasama, and K. Nishio. Onset of dynamic activity in globally coupled excitable and oscillatory units. *Phys. Rev. E*, 88(5):052907, 2013.

[160] L. M. Alonso and G. B. Mindlin. Average dynamics of a driven set of globally coupled excitable units. *Chaos*, 21(2):023102, 2011.

[161] D. Gomila, M. A. Matías, and P. Colet. Excitability mediated by localized structures in a dissipative nonlinear optical cavity. *Phys. Rev. Lett.*, 94(6):063905, 2005.

[162] H. Daido. Quasientrainment and slow relaxation in a population of oscillators with random and frustrated interactions. *Phys. Rev. Lett.*, 68(7):1073, 1992.

[163] H. Hong and S. H. Strogatz. Mean-field behavior in coupled oscillators with attractive and repulsive interactions. *Phys. Rev. E*, 85(5):056210, 2012.

[164] H. R. Wilson and J. D. Cowan. Excitatory and inhibitory interactions in localized populations of model neurons. *Biophys. J.*, 12(1):1, 1972.

[165] N. Brunel. Dynamics of sparsely connected networks of excitatory and inhibitory spiking neurons. *J. Comput. Neurosci.*, 8(3):183–208, 2000.

[166] M. San Miguel, V. M. Eguíluz, R. Toral, and K. Klemm. Binary and Multivariate Stochastic Models of Consensus Formation. *Comput. Sci. Eng.*, 7(6):67, 2005.

[167] M. Falcke, M. Bär, H. Engel, and M. Eiswirth. Traveling waves in the CO oxidation on Pt (110): Theory. *J. Chem. Phys.*, 97(6):4555, 1992.

[168] I. Schebesch and H. Engel. Wave propagation in heterogeneous excitable media. *Phys. Rev. E*, 57(4):3905, 1998.

[169] M. Hoyuelos, E. Hernández-García, P. Colet, and M. San Miguel. Dynamics of defects in the vector complex Ginzburg–Landau equation. *Physica D*, 174(1):176–197, 2003.

[170] A. I. Lavrova, S. Bagyan, T. Mair, M. J. B. Hauser, and L. Schimansky-Geier. Modeling of glycolytic wave propagation in an open spatial reactor with inhomogeneous substrate influx. *BioSystems*, 97(2):127–133, 2009.

[171] D. E. Postnov, F. Müller, R. B. Schuppner, and L. Schimansky-Geier. Dynamical structures in binary media of potassium-driven neurons. *Phys. Rev. E*, 80(3):031921, 2009.

[172] B. Sonnenschein, T. K. DM. Peron, F. A. Rodrigues, J. Kurths, and L. Schimansky-Geier. Collective dynamics in two populations of noisy oscillators with asymmetric interactions. *Phys. Rev. E*, 91(6):062910, 2015.

[173] D. Pazó. Thermodynamic limit of the first-order phase transition in the Kuramoto model. *Phys. Rev. E*, 72:046211, 2005.

[174] O. E. Omel'chenko and M. Wolfrum. Nonuniversal transitions to synchrony in the Sakaguchi-Kuramoto model. *Phys. Rev. Lett.*, 109(16):164101, 2012.

[175] H.-J. Wünsche, S. Bauer, J. Kreissl, O. Ushakov, N. Korneyev, F. Henneberger, E. Wille, H. Erzgräber, M. Peil, W. Elsäßer, and I. Fischer. Synchronization of delay-coupled oscillators: A study of semiconductor lasers. *Phys. Rev. Lett.*, 94:163901, 2005.

[176] J. Zamora-Munt, C. Masoller, J. Garcia-Ojalvo, and R. Roy. Crowd Synchrony and Quorum Sensing in Delay-Coupled Lasers. *Phys. Rev. Lett.*, 105(26):264101, 2010.

[177] E. A. Martens, S. Thutupalli, A. Fourrière, and O. Hallatschek. Chimera states in mechanical oscillator networks. *Proc. Natl. Acad. Sci. USA*, 110(26):10563–10567, 2013.

[178] K. Kostorz, R. W. Hölzel, and K. Krischer. Distributed coupling complexity in a weakly coupled oscillatory network with associative properties. *New J. Phys.*, 15(8):083010, 2013.

[179] M. R. Tinsley, S. Nkomo, and K. Showalter. Chimera and phase-cluster states in populations of coupled chemical oscillators. *Nature Phys.*, 8(9):662–665, 2012.

[180] J. F. Totz, R. Snari, Y. Desmond, M. R. Tinsley, H. Engel, and K. Showalter. Phase-lag synchronization in networks of coupled chemical oscillators. *Phys. Rev. E*, 92:022819, 2015.

[181] T. Gloveli, T. Dugladze, S. Saha, H. Monyer, U. Heinemann, R. D. Traub, M. A. Whittington, and E. H. Buhl. Differential involvement of oriens/pyramidale interneurones in hippocampal network oscillations in vitro. *J. Physiol.*, 562(1):131, 2005.

[182] K. T. Hofer, Á. Kandrács, I. Ulbert, I. Pál, C. Szabó, L. Héja, and L. Wittner. The hippocampal ca3 region can generate two distinct types of sharp wave-ripple complexes, in vitro. *Hippocampus*, 25(2):169, 2015.

[183] D. Cumin and C. P. Unsworth. Generalising the Kuramoto model for the study of neuronal synchronisation in the brain. *Physica D*, 226(2):181, 2007.

[184] C. Van Vreeswijk, L. F. Abbott, and G. B. Ermentrout. When inhibition not excitation synchronizes neural firing. *J. Comput. Neurosci.*, 1(4):313, 1994.

[185] J. Eccles. From electrical to chemical transmission in the central nervous system. *Notes Rec. R. Soc. Lond.*, 30:219, 1976.

[186] H. Sakaguchi and Y. Kuramoto. A Soluble Active Rotater Model Showing Phase Transitions via Mutual Entrainment. *Prog. Theor. Phys.*, 76(3):576, 1986.

[187] K. Wiesenfeld, P. Colet, and S. H. Strogatz. Synchronization transitions in a disordered Josephson series array. *Phys. Rev. Lett.*, 76(3):404–407, 1996.

[188] K. Wiesenfeld, P. Colet, and S. H. Strogatz. Frequency locking in Josephson arrays: connection with the Kuramoto model. *Phys. Rev. E*, 57(2):1563, 1998.

[189] R. Tönjes. Synchronization transition in the Kuramoto model with colored noise. *Phys. Rev. E*, 81(5):055201, 2010.

[190] J. De La Rocha, B. Doiron, E. Shea-Brown, K. Josić, and A. Reyes. Correlation between neural spike trains increases with firing rate. *Nature*, 448(7155):802–806, 2007.

[191] P. Ji, T. K. DM. Peron, F. A. Rodrigues, and J. Kurths. Low-dimensional behavior of Kuramoto model with inertia in complex networks. *Sci. Rep.*, 4(4783), 2014.

[192] P.-H. Chavanis. The Brownian mean field model. *Eur. Phys. J. B*, 87(5):120, 2014.

[193] Y. Kuramoto and D. Battogtokh. Coexistence of Coherence and Incoherence in Nonlocally Coupled Phase Oscillators. *Nonlinear Phenom. Complex Syst.*, 5(4):380–385, 2002.

[194] D. M. Abrams and S. H. Strogatz. Chimera states for coupled oscillators. *Phys. Rev. Lett.*, 93(17):174102, 2004.

[195] D. M. Abrams, R. Mirollo, S. H. Strogatz, and D. A. Wiley. Solvable model for chimera states of coupled oscillators. *Phys. Rev. Lett.*, 101(8):084103, 2008.

[196] C. R. Laing. Disorder-induced dynamics in a pair of coupled heterogeneous phase oscillator networks. *Chaos*, 22(4):043104, 2012.

[197] T. P. Vogels, K. Rajan, and L. F. Abbott. Neural network dynamics. *Annu. Rev. Neurosci.*, 28:357–376, 2005.

[198] J. G. Restrepo, E. Ott, and B. R. Hunt. Synchronization in large directed networks of coupled phase oscillators. *Chaos*, 16(1):015107, 2006.

[199] M. E. J. Newman, S. Forrest, and J. Balthrop. Email networks and the spread of computer viruses. *Phys. Rev. E*, 66(3):035101(R), 2002.

[200] M. Boguná and R. Pastor-Satorras. Epidemic spreading in correlated complex networks. *Phys. Rev. E*, 66(4):047104, 2002.

[201] C. Schmeltzer, A. H. Kihara, I. M. Sokolov, and S. Rüdiger. Degree Correlations Optimize Neuronal Network Sensitivity to Sub-Threshold Stimuli. *PloS one*, 10(6):e0121794, 2015.

Danksagung

In erster Linie möchte ich mich in ganz besonderem Maße bei meinem Betreuer und Mentor Prof. Lutz Schimansky-Geier bedanken. Seine anregende, ideen- sowie anekdotenreiche und zugleich unkomplizierte Art der Betreuung und Zusammenarbeit waren von immenser Bedeutung für das Erstellen dieser Arbeit. Genauso bedeutsam war seine andauernde Unterstützung bei organisatorischen Fragen. Vielen herzlichen Dank!

Ich danke der gesamten Arbeitsgruppe “Statistical Physics and Nonlinear Dynamics & Stochastic Processes” für die angenehme Arbeitsatmosphäre.

Großer Dank gebührt weiterhin allen Kollaboratoren. Ich danke Thomas Peron für die äußerst fruchtbare Zusammenarbeit und die unzähligen wissenschaftlichen und weniger wissenschaftlichen Diskussionen. Mit keinem anderen habe ich so viele Skype-Nachrichten ausgetauscht. Außerdem bedanke ich mich bei Prof. Francesc Sagués, Dr. Michael Zaks, Prof. Alexander Neiman, Prof. Francisco Rodrigues und Prof. Jürgen Kurths für die sehr gute Zusammenarbeit.

Für die vielen wissenschaftsnahen und oft darüber hinaus gehenden Gespräche während der Mittagessen und auch abseits davon, bedanke ich mich herzlich vor allem bei Paul, Arthur, Simon, Sebastian, Hamid, Justus, Felix, Christian, Júlio, David, Frau Rosengarten, Martin, Stefan, Patrick und Jörg.

Ich danke Sebastian, Hamid, Felix, Justus, Ines und Paul für das Korrekturlesen von Teilen dieser Arbeit.

Mein herzlichster Dank geht an Familie und Freunde, die mich unerschöpflich unterstützt haben und alles abseits der Wissenschaft in wertvollster Weise mit Leben füllen. Dabei möchte ich Ines, Souris, Sylvetta und Susi :) herausheben. Vielen Dank für alles!

List of publications

Parts of this thesis were published before in the following journal papers.

- Bernard Sonnenschein and Lutz Schimansky-Geier. Onset of synchronization in complex networks of noisy oscillators. Physical Review E, 85:051116, May 2012.
- Bernard Sonnenschein, Francesc Sagués and Lutz Schimansky-Geier. Networks of noisy oscillators with correlated degree and frequency dispersion. The European Physical Journal B, 86:12, Jan 2013.
- Bernard Sonnenschein, Michael A. Zaks, Alexander B. Neiman and Lutz Schimansky-Geier. Excitable elements controlled by noise and network structure. The European Physical Journal Special Topics, 222:2517, Oct 2013.
- Bernard Sonnenschein and Lutz Schimansky-Geier. Approximate solution to the stochastic Kuramoto model. Physical Review E, 88:052111, Nov 2013.
- Bernard Sonnenschein, Thomas K. DM. Peron, Francisco A. Rodrigues, Jürgen Kurths and Lutz Schimansky-Geier. Cooperative behavior between oscillatory and excitable units: the peculiar role of positive coupling-frequency correlations. The European Physical Journal B, 87:182, Aug 2014.
- Bernard Sonnenschein, Thomas K. DM. Peron, Francisco A. Rodrigues, Jürgen Kurths and Lutz Schimansky-Geier. Collective dynamics in two populations of noisy oscillators with asymmetric interactions. Physical Review E, 91:062910, Jun 2015.